I0885209

Enhancements and Limitations to ICT-Based Informal Language Learning:

Emerging Research and Opportunities

Rashad Ahmed
Miami University, USA

Abdu Al-kadi
Ibb University, Yemen

Trenton Hagar
Utrecht University, The Netherlands

A volume in the Advances in Linguistics and Communication Studies (ALCS) Book Series

IGI Global
DISSEMINATOR OF KNOWLEDGE

Published in the United States of America by
IGI Global
Information Science Reference (an imprint of IGI Global)
701 E. Chocolate Avenue
Hershey PA, USA 17033
Tel: 717-533-8845
Fax: 717-533-8661
E-mail: cust@igi-global.com
Web site: http://www.igi-global.com

Copyright © 2020 by IGI Global. All rights reserved. No part of this publication may be reproduced, stored or distributed in any form or by any means, electronic or mechanical, including photocopying, without written permission from the publisher.
Product or company names used in this set are for identification purposes only. Inclusion of the names of the products or companies does not indicate a claim of ownership by IGI Global of the trademark or registered trademark.

Library of Congress Cataloging-in-Publication Data

Names: Ahmed, Rashad, 1985- editor. | Al-kadi, Abdu, 1980- editor. | Hagar, Trenton, 1973- editor.
Title: Enhancements and limitations to ICT-based informal language learning: emerging research and opportunities / Rashad Ahmed, Abdu Al-kadi, and Trenton Hagar.
Description: Hershey, PA : Information Science Reference, 2020. | Includes bibliographical references and index. | Summary: "This book explores how learners perform ICT-based activities beyond classroom teaching. It also assesses the linguistic gains generated by informal ICT uses and suggests tools of gauging informal learning outcomes and possible effects on learners' proficiency"-- Provided by publisher.
Identifiers: LCCN 2019036936 (print) | LCCN 2019036937 (ebook) | ISBN 9781799821168 (hardcover) | ISBN 9781799821175 (paperback) | ISBN 9781799821182 (ebook)
Subjects: LCSH: English language--Study and teaching--Computer-assisted instruction for foreign speakers. | Mobile communication systems in education.
Classification: LCC PE1128.3 .E65 2020 (print) | LCC PE1128.3 (ebook) | DDC 428.0078/5--dc23
LC record available at https://lccn.loc.gov/2019036936
LC ebook record available at https://lccn.loc.gov/2019036937

This book is published in the IGI Global book series Advances in Linguistics and Communication Studies (ALCS) (ISSN: 2372-109X; eISSN: 2372-1111)

British Cataloguing in Publication Data
A Cataloguing in Publication record for this book is available from the British Library.

All work contributed to this book is new, previously-unpublished material.
The views expressed in this book are those of the authors, but not necessarily of the publisher.

For electronic access to this publication, please contact: eresources@igi-global.com.

Advances in Linguistics and Communication Studies (ALCS) Book Series

ISSN:2372-109X
EISSN:2372-1111

Editor-in-Chief: Abigail G. Scheg, Western Governors University, USA

MISSION

The scope of language and communication is constantly changing as society evolves, new modes of communication are developed through technological advancement, and novel words enter our lexicon as the result of cultural change. Understanding how we communicate and use language is crucial in all industries and updated research is necessary in order to promote further knowledge in this field.

The **Advances in Linguistics and Communication Studies (ALCS)** book series presents the latest research in diverse topics relating to language and communication. Interdisciplinary in its coverage, ALCS presents comprehensive research on the use of language and communication in various industries including business, education, government, and healthcare.

COVERAGE

- Interpersonal Communication
- Discourse Analysis
- Media and Public Communications
- Sociolinguistics
- Youth Language
- Language in the Media
- Computer-Mediated Communication
- Graphic Communications
- Computational Linguistics
- Cross-Cultural Communication

IGI Global is currently accepting manuscripts for publication within this series. To submit a proposal for a volume in this series, please contact our Acquisition Editors at Acquisitions@igi-global.com or visit: http://www.igi-global.com/publish/.

The Advances in Linguistics and Communication Studies (ALCS) Book Series (ISSN 2372-109X) is published by IGI Global, 701 E. Chocolate Avenue, Hershey, PA 17033-1240, USA, www.igi-global.com. This series is composed of titles available for purchase individually; each title is edited to be contextually exclusive from any other title within the series. For pricing and ordering information please visit http://www.igi-global.com/book-series/advances-linguistics-communication-studies/78950. Postmaster: Send all address changes to above address. Copyright © 2020 IGI Global. All rights, including translation in other languages reserved by the publisher. No part of this series may be reproduced or used in any form or by any means – graphics, electronic, or mechanical, including photocopying, recording, taping, or information and retrieval systems – without written permission from the publisher, except for non commercial, educational use, including classroom teaching purposes. The views expressed in this series are those of the authors, but not necessarily of IGI Global.

Titles in this Series

For a list of additional titles in this series, please visit:
https://www.igi-global.com/book-series/advances-linguistics-communication-studies/78950

Toward an Integrated Approach to Narrative Generation Emerging Research and Opportunities
Takashi Ogata (Iwate Prefectural University, Japan)
Information Science Reference • copyright 2020 • 432pp • H/C (ISBN: 9781522596936) • US $215.00 (our price)

Code-Switching for Equitable STEM Pedagogy Emerging Research and Opportunities
Miriam Chitiga (Fayetteville State University, USA)
Information Science Reference • copyright 2020 • 135pp • H/C (ISBN: 9781799809654) • US $125.00 (our price)

Applied Linguistics for Teachers of Culturally and Linguistically Diverse Learners
Nabat Erdogan (University of Central Missouri, USA) and Michael Wei (University of Missouri-Kansas City, USA)
Information Science Reference • copyright 2019 • 504pp • H/C (ISBN: 9781522584674) • US $195.00 (our price)

Language, Power, and Ideology in Political Writing Emerging Research and Opportunities
Önder Çakırtaş (Bingöl University, Turkey)
Information Science Reference • copyright 2019 • 206pp • H/C (ISBN: 9781522594444) • US $185.00 (our price)

Deep Semantics and the Evolution of New Scientific Theories and Discoveries
Tom Adi (The Readware Institute, USA) Hala Abdelghany (City University of New York, USA) and Kathy Adi (Readware Institute, USA)
Information Science Reference • copyright 2019 • 281pp • H/C (ISBN: 9781522580799) • US $185.00 (our price)

Argumentation and Appraisal in Parliamentary Discourse
Ernest Jakaza (Midlands State University, Zimbabwe)

For an entire list of titles in this series, please visit:
https://www.igi-global.com/book-series/advances-linguistics-communication-studies/7895

IGI Global
DISSEMINATOR OF KNOWLEDGE

701 East Chocolate Avenue, Hershey, PA 17033, USA
Tel: 717-533-8845 x100 • Fax: 717-533-8661
E-Mail: cust@igi-global.com • www.igi-global.com

Editorial Advisory Board

Muhammad Alasmari, *University of Memphis, USA*
Abdullah Alfaifi, *University of Indiana, Bloomington, USA*
Mohamed Daoud, *Carthage University, Tunisia*
Yusuf Demir, *Necmettin Erbakan University, Turkey*
Faiza Derbel, *Manouba University, Tunisia*
Cahit Erdem, *Afyon Kocatepe University, Turkey*
Christopher Hastings, *Southwest Tennessee Community College, USA*
Galip Kartal, *Necmettin Erbakan University, Turkey*
Mehmet Koçyiğit, *Afyon Kocatepe University, Turkey*
Tahar Labassi, *University of Tunis, Tunisia*
S. Mohanraj, *The English and Foreign Language University, India*
Gülçin Mutlu, *İzmir Democracy University, Turkey*
Mirella Silva, *University of Memphis, USA*
Christopher Williams, *Raytheon, USA*

List of Reviewers

Mayada Al-Maktary, *Manouba University, Tunisia*
Najeeb Al-Mansoob, *BAM University, India*
Hussein Almaktary, *ISLAB, Tunisia*
Mohialdeen Alotumi, *Sana'a University, Yemen*
Taha Hezam, *Bisha University, Saudi Arabia*
Ferit Kılıçkaya, *Burdur Mehmet Akif Ersoy University, Turkey*
Saadet Korucu-Kis, *Necmettin Erbakan University, Turkey*

Xinghua Liu, *Shanghai Jiao Tong University, China*
Asma Maaoui, *University of Tunis, Tunisia*
Emine Şendurur, *Ondokuz Mayıs University, Turkey*
Amr Yassin, *Universiti Kebangsaan Malaysia, Malaysia*

Table of Contents

Foreword

Information and communication technology (ICT), formerly referred to as information technology (IT), is now common as digital technology. It provides access to information through telecommunication. Although ICT is similar to IT, its primary focus is on communicative technology. Synchronous and asynchronous communication modes have provided food for thought in academia for years. The omnipresence of modern ICTs including the Internet, cell phones, social media, and other communicative channels have created new conditions and environments to learn foreign languages quickly and easily inside and outside the classroom. The book "Enhancement and Limitations to ICT-based Informal Language Learning" is the product of authors, editors, reviewers, and advisory editorial board who are all well-suited to tackle its subject-matter – as they are educators with myriad international experiences in the field of learning and teaching English as a second language.

This publication draws chiefly on two learning paradigms: technology-enhanced learning (TELL) and informal language learning. It overthrows the hegemony of formalized learning and hails informal learning as a patent part of humankind, while also focusing on informal learning under the aegis of digital technologies. It accounts for the implications of shortening the distance between formal and informal learning of English as a second language (L2). In fact, using ICT in formal language learning is no novelty as it has been in use since the inception of early CALL in the 1960s – the period during which technologies were employed in systematic and planned ways to facilitate learning mechanism in formal education. Later on, advances in digital ICTs gave way to new learning settings, opportunities, strategies, and inducement. Advances in learning and teaching paradigms and technological innovations reciprocally allowed social and collaborative learning in face-to-face settings as well as in non-administered situations, largely in virtual spaces. This book is centered on this latter learning mode. It rests on the merits of informal language learning and it minimizes the hoopla that surrounds classroom learning. It builds–through digital technology– a bridge between formal and informal learning. The mission of the book is to support a balanced and flexible learning environment in lieu of the long-established formal education confinements.

The book contains two sections: theoretical accounts and empirical studies. Whereas the former comprises the first three chapters that lay the theoretical background of the book, the latter compiles evidence of learning with ICT in informal situations. All chapters were written with an awareness that they will be especially useful for pedagogues, researchers and teachers who are interested in L2 learning and teaching, CALL, TESOL, and education. The chapters have profound implications for L2 pedagogy (English is a working example). They exhibit how learners manipulate accessible ICTs to enhance their learning of English beyond the traditional classroom. By suggesting how to apply methods and techniques of learning language in an informal manner, the audience of the book will be able to enhance formal English by making informal decisions and better choices on the basis of the research findings delineated in the chapters.

The book is indeed an important addition to the existing scholarly works in the area of TELL. One important issue the book has taken up is that the advent of digital technology has given English a new status. English has become the mother-tongue of online communities because many ICT users communicate through English. That is, English is not exclusively taught through books in schools and universities to non-English speakers. The traditional classification of English as English as a Foreign Language (EFL) and English as a Second Language (ESL) has been subsided partially due to this worldwide electronic usage of the language. By using English online or through ICT applications or games, users are exposed to varieties of English, not only the standard British or American varieties. World Englishes and English as a lingua franca (ELF) have become common in virtual spaces where learners spend much of their time connected to technological innovations. Hence, it seems awkward to stick to the old classifications of English when discussing new technological advances that have facilitated and shaped new approaches to language learning and teaching, and –overall– new ways of thinking. Furthermore, the book brings to the foreground a multiple-method approach to assess informal language intake. It pinpoints that language assessment formats such as multiple-choice questions and matching activities in the form of quizzes, midterms or final examinations are less suitable to evaluate informal language learning that results from spontaneous uses of ICTs. Alternative assessment tools may be adopted instead with a range of evaluation tools: quizzes, concept mapping, peer review, portfolios, performance assessment, reflective writing and media creation, rubrics, interviews, observations, and self-reporting. These alternative assessment tools enable learners to have a say in what they learn. They become able to determine their learning objectives independently from formal authorities, entrenching learning with elements of autonomy and learner-centeredness.

Touching on the above, the book presents highly-readable and well-organized, insightful ideas to re-orient English language teaching in the international context with a primary focus on ELF rather than EFL. The volume provides implications for re-adjusting learning settings, broadening the learning ecology, and embracing new learning venues, methods, strategies, and assessment tools. As ICT makes language learning possible anywhere at any time, the common term of 'classroom' is replaced by learning spaces. In essence, the book instills self-governing learning elements into the language learners to be linked with what they learn formally. The contributors hold the view that learners should be given an opportunity to determine what and how to learn: to learn on their own, monitor and assess their learning. Only when formal and informal learning activities are pedagogically interlined can we capitalize on informal learning significance.

Abraham Panavelil
University of Nizwa, Oman

Preface

Digital technology dominates all spheres of modern life making language learning boundless from the classroom due to the abundance of information and resources available online. The burst of information and communication technology (ICT) applications, search engines, and Massive Open Online Courses (MOOCs) have made individuals grumble about lack of time to access this heap of information, which was one day a rarity. ICT has created improvised sites of learning, making language learning cross the boundaries of formal education. Informal learning experiences have been discarded for years, but the advent of modern ICTs has accelerated chances of self-managed learning, making it no longer seen as an inferior form of learning. A MOOC provides a partial solution for information seekers, and it seems to be perfect as it facilitates obtaining the desired information, knowledge and skills in one place in lieu of surfing the World Wide Web to select from an ocean of information.

MOOCs are examples of many other ventures that have made informal learning commonplace. This ICT boom has proliferated ground for researchers and educators to recognize informal learning as it amends, supplements, and complements formal learning scenarios. It also potentially compensates possible shortcomings of formal education (Godwin-Jones, 2018; Jarvis, 2014). Notwithstanding diversity of definitions, both learning modes overlap. Formal learning includes elements of informal learning, and informal learning, likewise, has elements of formality. This reciprocal relationship blurs the distinction between formal and informal learning. In the realm of second language (L2) learning and acquisition, it is generally accepted that ICT boosts L2 in formal settings and offers spaces to facilitate active, collaborative, creative, integrative, and evaluative learning. These spaces supersede the traditional learning venues called ‘classrooms’ and can be seen as an advantage over the traditional language learning.

Researchers have looked into how wired and wireless ICTs– dubbed desktop and non-desktop technologies– have extended the territories of learning. While desktop ICTs (wired technologies) are mostly attached to formal learning as they are well-situated in formal settings, wireless ICTs (non-desktop) are associated with informal learning. Ubiquitous media technologies (mobile devices) have facilitated the mobility of learning – crossing classroom borders while making learning more flexible. The bulk of research on technology-enhanced language learning (TELL) is based on structured learning with uses developed by instructors and/or researchers. However, the time and efforts spent on TELL beyond formal instruction continues to be overlooked. ICT, which creates a wide range of electronic possibilities, surrounds learners with knowledge sources and resources, "varying from simple electronic dictionaries to full-fledged online language programs" (Suvorov& Hegelheime, 2013).

Disregarding this technological penetration results in food for thought that this book brings to the foreground. With students nowadays having electronic devices (i.e. laptops, notebooks, iPods, and smartphones), accessing online resources, tutorials, and learning materials has become second nature to the majority who pass much of their time using such devices. In addition, learners themselves construct knowledge by feeding virtual spaces with their postings, selfies, comments on others' posts, audio/video conferencing, and the like. These activities are mostly informal practices which have been neglected; and only quite recently have been acknowledged, skeptically though, as learning opportunities that enhance formal learning. When it comes to using the English language, learners are exposed to and may acquire varieties of English – not only the standard varieties. When assessing learners' performance via standard tests, this informally-obtained English uptake is not taken into account in most ESL and EFL situations as it is difficult to gauge informal learning outcomes separately from formal tests. Equally, it is impractical to measure informal learning via formal tests.

The book is designed with certain models and theories in mind; models such as Mobile-Blended Collaborative Learning model, Technology Acceptance Model (TAM) (Davis, Bagozzi, Warshaw, 1989), use of technology model (UTAUT) (Venkatesh, Thong & Xu, 2016), informal language models (Rogers, 2004; Sockett, 2014), and theories such as SLA theories (Ellis, 2005; Krashen, 1982; Mitchell & Myles, 1998), and experiential learning theory (Kolb, 1984). The publication connects unbridged formal and informal language learning. It reports data from L2 contexts, covering topics of technology acceptance for informal language learning and practices of this informal language uptake. It encompasses contributions from ten authors from Malaysia, Saudi Arabia, Yemen, Turkey and the USA as well as fifteen reviewers who are educators, language teachers, and researchers. All hold a PhD or MA in TESOL, Applied linguistics, education, and relevant branches of knowledge.

A salient point the book brings into the fore is a dissatisfaction with classifying English language teaching as EFL and ESL. This terminology has become outdated because it was determined on the basis of colonial factors. One reason for this obsolescence is that English is the mother-tongue of computers and allied devices. As such, the significance, status, and methods for teaching and learning English have changed. ICT users reside in a community imbued with World Englishes – which has been a scarcity in EFL settings. Unlike EFL, ESL contexts take up English for official use. However, in the context of modern technology, English has gained a new status with different needs and significance. The linguistic map of English in the worldwide context has changed, bringing about a new classification of English – the terms English as an International Language (EIL), English as a Lingua Franca (ELF), and World Englishes have surfaced and vigorously gained currency. In other words, English is the lingua franca of the world and is shared by the worldwide community.

One more point the book touches on is germane to what Krashen termed, *second language learning* and *language acquisition*. Giving the difference between the two concepts, for L2 acquisition to take place there should be ample comprehensible input, low affective filter, and sufficient exposure to the target language – all are possible in TELL situations wherein learners are exposed to English everywhere around the clock. Also, L2 learners can get timely feedback from other users (Khaddage, Lattemann, & Bray, 2011) because TELL creates a productive environment that facilitates language learning and may promote Krashen's concept of 'language acquisition'. Building on Krashen's hypothesis of comprehensible input, what happens in the classroom is *language learning;* for acquisition to occur, it should be complemented with out-of-class language practices. The latter (out-of-class use) differs from learner to learner and from context to context. The context of ICT looms to be a global context where a global community of English users contact through the world via a lingua franca or World English.

The main thrust of this publication is bridging the gap between formal and informal language learning with the help of digital technology. The book delves into ICT-enabled opportunities that today's learners have when they learn informally. It explores –in a cogent manner– how learners perform ICT-based activities beyond classroom teaching, assesses the linguistic gains generated by informal ICT uses, discusses possible effects on learners' proficiency, and suggests tools for gauging informal learning outcomes. It also elucidates limitations that impede recognizing this mode of learning in the worldwide context. The book is meant for pedagogues, researchers, and teachers who are interested in second language learning and teaching, computer-assisted language learning (CALL), and TESOL. It has profound implications as it exhibits how learners manipulate accessible ICTs to enhance their English learning beyond the confinements of the traditional classroom. It also contributes to the development of new courseware directed towards ICT-based language learning. By suggesting how to apply methods and techniques of learning language informally, stakeholders can promote formal English instruction by making informed decisions and better choices on the basis of research findings rather than intuitions.

THE AUTHORS

The contributors to this volume represent a blend of scholarly acumen and practical experience.

These authors have contributed to this volume based on their teaching, research background, and familiarity with technology-based education in L2 contexts. Perhaps the most important qualification these authors share is a commitment to shift functions and forms of learning away from the conventional view of learning dependent on classroom English. They have a rich understanding of the processes involved in language acquisition and teaching as well as the subject matter and pedagogical preparedness to plan, develop, and deliver meaningful and engaging lectures and effectively manage student-centered classes.

ORGANIZATION OF THE BOOK

The book is a joint cooperation between the editors, editorial advisory board (EAB), authors, and reviewers. The EAB are experienced teachers, authors, and senior practitioners who are affiliated with universities in Tunisia, India, Turkey, and the USA. The volume is organized into eight chapters, bringing in experiences, practices, and insights from authors in Malaysia, Saudi Arabia, Yemen, Turkey, and the USA. The chapters were reviewed by experienced professionals and are classified into two parts: theoretical analyses of the topic and empirical studies. Besides theoretical accounts, the book contains data collected from learners and educators, results of quantitative and qualitative research projects, and condensed perceptional data, cross sectional studies, involving surveys, interviews, and case studies. A brief description of each chapter follows.

Section 1: Theoretical Accounts

Chapter 1: Informal Language Practices via Digital Technology – The Neglected Proficiency

In this chapter, Dr. Abdu Al-Kadi discusses numerous ICT ventures which afford second language (L2) learners opportunities to practice the target language on their own at their pace. The author argues that the linguistic gains that foreign language learners obtain by using digital technology on their own is not taken into account when it comes to evaluating their overall linguistic competence; it becomes a wasted proficiency. Proficiency, in this chapter, defined not in terms of the native speakers' model but in terms of the World Englishes proficiencies – proficiencies which include the growing varieties of English under the auspices of modern information and communication technologies such as internetese, emailese, textese that language pundits such as the scholar David Crystal believes are an addition to the standard variety of English. The chapter addresses the most relevant assumptions and research on SLA, CALL, and informal learning. It clearly lays out the issues and trends related to language learning and ICT. It also touches on major concerns when trying to evaluate and assess such vital characteristics of today's SLA environments.

Chapter 2: A Guide for Mobile-Assisted Language Learning in Informal Settings – Pedagogical and Design Perspectives

In this chapter, Dr. Emine Şendurur covered a number of technological transformations and breakthroughs that have created a paradigm shift in terms of how users and creators access, design and consume educational materials in the modern world. The author discusses 21st century trends requiring certain skill sets ranging from collaboration to creative thinking. She outlines the multiple pathways, tools and issues facing modern educational practitioners and learners, touching on how curriculum design is impacted by these changes and shows how the transformational nature of technology will necessarily change the design of curricula going forward. This is particularly important since the author makes efforts to discuss the impact of assessments through new technological pathways. The author states that the new Information Age imparts new demands on the learner that set it apart from the Industrial Age before it (apart from moving from encyclopedias to online tools). The chapter closes on a note of guidance on MALL.

Chapter 3: Impediments to Gauging ICT-Based Informal English Learning Outcomes

In this chapter, Dr. Taha Hezam, reporting from the Saudi EFL context, has chosen a challenging topic - informal learning and informal assessment. Both are nebulous; and inadequate help is available from previous research. The chapter elucidated major challenges of measuring informal second language learning through digital technology. The author alleged that the assessment can happen informally through reflective tasks. Reflection which can be guided or monitored can help obtain adequate information on both informal learning and self-assessment of informal learning. Adults have a fair idea of their learning the first language (which is informal) and they have a reasonably correct measure of the competence achieved. But as the author rightly says, this is difficult to capture, hence a bit of monitoring by providing a protocol can be helpful. It clearly is directed at language learners, educators and researchers. Its main focus is ICT and informal learning strategies.

The chapter adequately articulates a problem and its implications. The author argues that because of the fuzzy nature of the language construct, informal learning assessment may be equated to formative assessment including diagnostic testing and other informal assessments that classroom teachers usually use to gauge their learners' abilities in certain areas.

Section 2: Empirical Evidence of Informal Learning

Chapter 4: Learner-Initiated Language Learning Through Social Media Sites

The rise of modern technology among learners of English has received the attention of scholars in the field of English language teaching and substantial research exists on this issue. However, the specific language benefits of SMSs in informal settings have not been sufficiently explored. This omission served as the primary rationale for this chapter in which Dr. Rashad Ahmed presents the results of a study on the practices and perceived benefits of using social media sites (SMSs) in English to supplement formal education. Specifically, the study focused on investigating the use of three SMSs (Facebook, WhatsApp, and Twitter) and the language-learning benefits gained as a result of using these platforms in informal settings. Based on the findings, it is implied that informal uses of SMSs can be supplementary to the usual classroom procedures, mainly in providing authentic contexts for learning, alternative assessment, and facilitating communication between students and teachers.

Chapter 5: Adult Language Learners' Informal Employment of ICT Applications and Websites to Assess Their English Skills

In this chapter, Dr. Ferit Kılıçkaya argues that assessment informs stakeholders of not only learners' performance but also whether objectives set for the curricula are successfully reached. More importantly, the main function of assessment is to determine and improve student learning and to act accordingly. The author discusses the recent concept of informal assessment of language proficiency using technology. The chapter creates a context for discussing the topic appropriately. The author had data from a large sample, which was used efficiently to draw inferences. It also lists free various websites and ICT applications available for practicing and assessing individuals' own language competence. The author

Chapter 6: An Investigation Into EFL Pre-Service Teachers' Out-of-Class Self-Regulated Learning Experiences With ICT

In this chapter, Dr. Saadet Korucu-Kis joins the efforts of other researchers who suggest that supporting learners to self-direct their language learning beyond formal contexts is very important to exploit the potential of technology for language learning gains. This issue is of particular relevance to countries, where English is learnt and taught as a foreign language and learners' exposure to the target language is limited to classroom teaching. The author explores learners' use of technology outside the classroom and documents sites which are useful for them as well as challenges they face when using such sites. The author highlights that the current practices are unbridged with the mainstream education, bringing to the core ideas for future research.

Chapter 7: EFL Learning Beyond the Wall With MALL – College Students' Perceptions

Recent technological developments characterized by mobility, connectivity, and affordability provided impetus for a plethora of research in L2 context. This motivated Dr. Mohaildeen Al-Otomi to discuss the acceptance of MALL for language learning in the Yemeni context. The author methodically guides the reader through the goals, history, current paradigms and, ultimately, toward his own study. The author argues that an instrument branded taboo in the classroom has now become a facilitator. The topic is linked to ICT and shown as an evolutionary product. The chapter examines perceptional evidence of 150 learners of English and their use of MALL for informal EFL learning. In addition, the study discusses the integration of MALL into EFL instruction through allowing, guiding, and encouraging learners to harness their mobile phones in ways that serve the pedagogical objectives of EFL lessons.

Chapter 8: Formalizing Informal CALL in Learning English Language Skills

Which ICT skills do learners need to formalize informal CALL? What are the learning strategies that learners use to study informally through CALL? Which CALL elements do learners need to formalize informal CALL? Professor Norizan Abdul Razak, Dr. Tengku Nor Rizan, and Dr. Amr Yassin examine these questions and present background information on Formal vs. Informal Learning and CALL vs. Informal CALL. In addition, the chapter provides insight into how to formalize informal CALL in learning English language skills. The discussion, based on a sample of learners who experienced informal CALL, illustrates the issues, problems, and trends related to the theme of this book.

In brief, the book contributes to the field of second language learning and teaching, digital learning, and education. It discusses digital technology as a timely revolution to help language learning transcend conventional classroom teaching. It is an overview of what is hoped to bridge classroom practices with those practices that learners continue performing after formal structured lessons. This is a means of long-life, self-directed learning that does not end at the walls of the classroom. Another contribution is related to autonomous learning, which is a desire of modern education in making learners inclined to learn on their own (i.e. independent from formal agencies). Despite this, digital technology does not always work flawlessly. Learning informally without having a framework, models, or benchmarks makes it difficult to measure the outcomes of informal language intake. Additionally, in many contexts the idea of autonomous learning and online certification is quite unconvincing. Because technological innovations are changeable, the outcome is unpredictable and this perhaps leaves policymakers and pedagogues with a second thought when considering informal learning in users' overall proficiency. While it is important to keep abreast of the latest development in CALL, MALL, and their associated applications, it is reasonable to caution against overly optimistic expectations from informal uses of technology. For one thing, although ICT is bound to have profound effects on L2 education, free and uncontrolled ICT informal uses "involve the messy, unpredictable use of human language for motivated, authentic purposes, a phenomenon that does not lend itself to laboratory controls" (Meskill & Quah, 2013, p. 41). Finally, technology should be touted as neither a panacea nor as a downfall. It presents benefits and challenges for learners and learning institutions, and the book touches on this thoroughly.

Rashad Ahmed
Miami University, USA

Abdu Al-Kadi
Ibb University, Yemen

Trenton Hagar
Utrecht University, The Netherlands

REFERENCES

Davis, F., Bagozzi, R., & Warshaw, P. (1989). User acceptance of computer technology: A comparison of two theoretical models. *Management Science*, *35*(8), 982–1003. doi:10.1287/mnsc.35.8.982

Ellis, R. (2005). Measuring implicit and explicit knowledge of a second language: A psychometric study. *Studies in Second Language Acquisition*, *27*(2), 141–172. doi:10.1017/S0272263105050096

Khaddage, F., Lattemann, C., & Bray, E. (2011). Mobile apps integration for teaching and learning. (Are Teachers Ready to Re-blend?). In M. Koehler & P. Mishra (Eds.), *Proceedings of Society for Information Technology & Teacher Education International Conference 2011* (pp. 2545-2552). Chesapeake, VA: AACE.

Kolb, D. A. (1984). *Experiential learning: Experience as the source of learning and development*. Prentice-Hall.

Krashen, S. (1982). *Principles and practice in second language acquisition*. Pergamon Press.

Meskill, C., & Quah, J. (2013). Researching language learning in the age of social media. In M. Thomas, H. Reinders, & M. Warschauer (Eds.), *Contemporary computer-assisted language learning* (pp. 39–54). London: Bloomsbury Academic.

Mitchell, R., & Myles, F. (1998). *Second language learning theories*. Oxford, UK: OUP.

Rogers, A. (2004). *Looking again at non-formal and informal education - towards a new paradigm, the encyclopedia of informal education*. Retrieved from www.infed.org/biblio/non_formal_paradigm.htm

Sockett, G. (2014). *The online informal learning of English*. London: Palgrave Macmillan. doi:10.1057/9781137414885

Suvorov, R., & Hegelheime, V. (2013). Computer-assisted language testing. In A. J. Kunnan (Ed.), *The companion to language assessment* (pp. 594–613). Hoboken, NJ: John Wiley & Sons, Inc. doi:10.1002/9781118411360.wbcla083

Venkatesh, V., Thong, J., & Xu, X. (2016). Unified theory of acceptance and use of technology: A synthesis and the road ahead. *Journal of the Association for Information Systems*, *17*(5), 328–376. doi:10.17705/1jais.00428

Acknowledgment

We would like to acknowledge the help of all the people involved in this project and, more specifically, to the authors and reviewers. Without their support, this book would not have become a reality. Each one deserves our sincere gratitude for their contributions and endurance. Similarly, we wish to acknowledge the valuable contributions of the reviewers regarding the improvement of quality, coherence, and content presentation of the chapters. Most of the authors also served as reviewers; we highly appreciate their double task.

We are also obliged to the erudite Advisory Editorial Board members whose remarks and evaluation supported us in getting this publication off thc ground, steering the volume along and bringing it to fruition.

Alike, we cannot forget to render thanks to Miss Halle Frisco, Ms. Jan Travers, from IGI Global, for their support throughout conducting the project.

Introduction

As documented by Kennedy, Lathan and Jacinto (2016), students and teachers alike are surrounded nowadays with technology and make extensive use of it. Engagement with learning opportunities beyond the classroom and the workplace is common practice across age groups (Prensky, 2001). For instance, teachers intentionally join expert online communities to keep updated and share ideas, concerns and experiences with multiple others. Young learners especially are immersed in natural, autonomous and self-determined use of online resources (Hannibal & Jensen, 2019). Options and affordances of emerging information and communication technologies (ICTs), especially under Web 2.0 and Web 3.0 (Chun, 2007; Goodwin-Jones, 2005; 2011; Thomas 2009), entails the intentional and incidental use of interactive and participatory learning tools, social networks, learning management systems (LMSs), multimedia tools and products reflecting wide-ranging styles of language use. Academic researchers have been tapping into this phenomenon among young school-age children (e.g. Sundqvist, 2009; Sundqvist, 2015; Tar, 2017) and identified what young people do when online. At the top of the list there is listening to music, playing games, watching videos, reading newspapers online, and interacting with peers. If not harnessed in formal education settings, the wealth of materials and interpersonal interaction activities are noted by these researchers as missed opportunities for 'uptake' and development of digital learning strategies.

Teachers, or at least the enthusiasts among them, have been developing their own blend of technology-aided instruction and, by doing so, been shaping and transforming their learners' experiences (Levy, 2009; Webb & Cox, 2004). For instance, when they assign students a class paper or project and require the use of internet resources, teachers reorient the students' random navigation habits towards a learning purpose and may fashion the assignment so that a classroom-related task is connected to specific real-life situations and positions learners as active participants in the process of completing the task (George & Sanders, 2017). The teacher can also orient the task so that work "outside the classroom" is part of the process. This way, learners are required to identify, access, evaluate, and analyze digital resources and, depending on the complexity of the task, they may be able to reach higher order skills of creating a story, an audio or video documentary, or writing a succinct summary to be shared with their peers (Amer, 2006; Zhao, 2012). This type of ICT-supported integrative pedagogy encourages learners to navigate through formal and informal settings as well as understand how the classroom and the outside world may be interconnected (Dubreil & Thorne, 2017; Lai, Zhu, & Gong, 2015).

English language teachers, in particular, encourage learners to engage in informal online communication, to use multimedia resources outside the classroom and involve their students in e-collaboration projects as good substitutes for the lack of exposure to authentic language use in contexts where exposure to the language is limited. A case in point is the Cultura project which carefully weaved e-collaboration into the formal language learning situations across contexts and built-in a research component to gauge the outcomes (Furstenberg, Levet, English & Maillet, 2001). Work in this area has been sustained over the last few decades to encompass pedagogical goals like developing intercultural competence (Byram, 1997), cultural understanding and cross-cultural awareness for global citizenship (O'Dowd, 2018). Thus, pedagogically-oriented CALL is teacher-led as any learner involvement with technology out of the classroom is plotted and controlled by the teachers as designers. However, by incorporating out-of-class activities into the schemes of conventional classroom work opened teachers' and learners' eyes to the benefits of informal learning (Cook and Smith, 2004). Teachers witnessed firsthand better learner engagement, a boost in motivation and enhancements in the quantity and quality of learner language production. It is this type of engagement with technology in schools which placed learners and teachers at the crossroads between formal and informal learning environments thus fostering a view of formal and informal learning/teaching modes as being intertwined and complimentary (Chun, 2007, Dubreil & Thorne, 2017; Reinders & Benson, 2017).

As ICTs have become omnipresent and hardware and software more and more user-friendly, self-access to knowledge has become a natural choice as people realize that there is a world of knowledge available at their fingertips. User-friendly devices meant that they could use the technology autonomously for personal, professional and academic purposes or to perform mundane tasks like booking a flight or making a cake (Selwyn, 2013). The needs for formal and informal learning may well be distinct but there is need for understanding the learning processes in self-access mode and within informal settings. While the two worlds of "formal" and "informal" learning have been kept separate within systems of education and training in most societies, the boundaries have been blurred between the two and that questions about the increasing impact of informal learning with technology have become pressing questions in the global 21st century (Lai, 2017; Little & Thorne, 2017). Opportunities for informal learning "in the digital wilds," are abundant, varied, open-ended but somewhat chaotic, and therefore, it is for language educators to ponder upon ways to address the complexity of the situation and not leave learners to their own devices (Sauro and Zourou, 2019, p. 5).

The current book, *Enhancements and Limitations to ICT-Based Informal Language Learning*, is then a good starting point on the path to developing a vision for technology-supported learning/ teaching in informal settings. Indeed, the contributions in the book report on instances of informal and self-directed technology-use situations and explore ways to measure the possible gains in learner proficiency. The contributors in this book highlight the view that informal learning is a continuing process of reconstruction depending on the goals users assign to the learning opportunities available to them. The different contributors to this book investigated situations from the learner/user perspective, providing snapshots of the experiences of students in different tertiary level institutions from around the world. Put together, the chapters provide an overview, though somewhat restricted in time and space, of this emerging mode of learning and document the options and challenges users experience along the way. A major challenge which transpires across the chapters is the demarcation between 'formal' and 'informal' education and a sense of loss of opportunity for better quality learning if language educators do not bridge the gap between them.

This collection is undoubtedly a valuable addition to the scarcity of studies exploring learner experiences within 'informal learning' across contexts such as Hyland's (2004) in Hong Kong, Lee (2018) in Korea, Sundqvist (2009 and 2015) in Sweden, and Tar (2017) in Finland and Vietnam. Certainly, the focus on informal learning is still in its infancy and more studies are needed to identify the learning resources, spaces and learner actions in the process. In this sense, the book is timely and highly significant. If informal education is to be embraced as a complementary mode or an alternative, what can be the enhancements and challenges to this new teaching mode? The book's numerous chapters, it is hoped, will produce a cumulative effect and contribute to an understanding of the manifestations of this on-going phenomenon as well as the challenges and impediments encountered by learners.

Organization of the Book

The contributions to this book are grouped under a theoretical part (theoretical accounts) and an empirical part (pedagogical applications). All in all, eight chapters deal with the central theme of informal language learning with technology. Each of the chapters is concerned with exploring aspects of the conceptualization of formal and informal learning and how they may operate in real educational settings. The chapters help build the case for the potential value and feasibility of informal learning and to determine the differing perceptions of and attitudes towards learning beyond the classroom as inferred from the data obtained from students in different institutions. The book also seeks to find evidence for whether, and to what extent, informal learning as supplement or stand-alone can result in gains in learner proficiency. Thus, the contributions explore 'bridging the gap' between the two modes of learning while others focus on self-access, self-regulation, and self-assessment. Chapter 1 juxtaposes these two modes of learning, presents a thorough analysis of informality and critically examines the widely-held views of informal learning. This chapter can be regarded as the theoretical foundation of the book. The author situates informal learning of English in the context of English as the language of the world, the language of technology and of intercultural exchanges and discusses the pedagogical concerns over the relevance of uptake from informal learning opportunities for L2 development. The important educational question asked in Chapter 1 is whether we are to leave students unprepared for participation in Web 3.0 and Web.4.0 environments or 'formalize informal learning'. The answer is not straightforward and, while the studies cited in the chapter are inspiring, the author cautions against the unprincipled design, implementation or evaluation of the outcomes of formal/ informal blends. Chapter 2 connects with Chapter 1 by discussing the possibility of implementing ICT-based informal learning at the level of design, selection, and learner involvement. The author underscores the potential role of mobile technology applications in enriching learning experiences beyond the classroom. Chapter 3 raises the issue of self-testing with technology and the dilemma of wanting to measure the impact of informal technology-based learning activities using alternative ways of assessing performance which are compatible with a learner-centred pedagogy.

Five research studies are included in the second part of the book, the aim of which is to present information on the experience of learning informally and pinpoint the impact of engaging in self-selected and self-regulated informal online learning experiences. Chapters 4, 5, 6, and 7 are exploratory in nature and tell the story from the perspective of the students. Chapter 4 deals with the students' perception of language-related benefits they gained from accessing content from social media sites and from participation in social media discussions Chapter 5 explores which applications adults use to assess themselves and how they use them. Chapter 6 explores to what extent self-regulated learning opportunities out-of-class settings are used by pre-service teachers and what self-regulation processes are triggered by such use. Chapter 7 focuses on the case of Mobile-assisted language learning (MALL) among Yemeni college students in an attempt to gauge the extent to which they accept the use of mobiles to learn English informally. Chapter 8, adopting a hermeneutical phenomenological approach, delves into formalizing informal CALL in learning English language skills. The chapter highlighted the need to integrate the elements of behaviourist CALL, cognitive CALL, constructivist CALL and socio-constructivist CALL to make informal CALL more effective. Through these chapters, the editors intend to provide evidence as to whether students engage with ICT-supported informal learning and whether they perceive these experiences as consolidating their existing language proficiency level and developing their language skills.

Significance of this Book

The declared aim of the book is to understand the situation of informal learning and to explore how learners perceive that particular mode of learning. The book gives snapshots of informal learning situations and experiences and provides through the contributions, insight into the issues surrounding learning in this mode. The editors embrace the position that informal learning is worthwhile when combined with formal instruction or added on by the learner in self-access format. The chapters in the first part of the book especially highlight the fact that a conceptualization of "informal learning" is necessary while the empirical chapters build the case for informal learning as an individualized choice. As such the compilation of nine chapters represents an invaluable resource for teachers, teacher educators, and administrators, about informal learning and the perspectives of self-directed learners as users. Readers of the book are in a better position to decide which of the various contributions can inspire them to better manage the formal/informal interface of language learning in their own context. The first part of the book provides guidance to practitioners who are intrigued by their learners' abundant use of technology out of the classroom. These chapters provide, in part, answers to the pedagogical questions including what procedures and strategies are appropriate for the open-ended and unpredictable nature of informal learning and whether informal learning experiences can be combined with pre-existing knowledge or reconciled with prescribed curricula specifications found in formal school settings.

CONCLUSION

On the whole, and as explained earlier in this chapter, this book calls our attention to informal learning as a natural and inevitable outcome of the technological affordances and availability of resources. It contributes to our understanding of its potential for connecting the classroom and the real world (Lai, Zhun & Gu, 2015). It focuses on the learners' connections with the sources of knowledge available and how the connections form and transform learning experiences and outcomes (OECD, 2015). Certainly, the de facto situation of learning 'in the wilds' does challenge educators, educational institutions and societies in the present century. By putting the issue on the table, the editors of the book have triggered thinking about the formal/informal interface and called attention to informal learning as being the learning mode of the future whereby informal learning may not only mean rethinking educational practice, it may also signal the decline of the existing dominant modes of centralized education, packaged curricula and teacher-centred pedagogy (Sauro & Zourou, 2019; Selwyn, 2013).

Faiza Derbel
University of Manouba, Tunisia

REFERENCES

Amer, A. (2006). Reflections on Bloom's revised taxonomy. *Electronic Journal of Research in Educational Psychology*, *4*(1), 213–230.

Byram, M. (1997). *Teaching and assessing intercultural communicative competence*. Clevedon, UK: Multilingual Matters.

Chun, D. (2007). Come ride the wave: But where is it taking us? *CALICO Journal*, *24*(2), 239–252. doi:10.1558/cj.v24i2.239-252

Cook, J., & Smith, M. (2004). Beyond formal learning: Informal community eLearning. *Computers and Education*, *43*(1-2), 35-47.

Dubreil, S., & Thorne, S. L. (2017). Social pedagogies and entwining language with the world. In S. Dubreil & S. L. Thorne (Eds.), *Engaging the world: Social pedagogies and language learning* (pp. 1–11). Boston, MA: Cengage.

Furstenberg, G., Levet, S., English, K., & Maillet, K. (2001). Giving a virtual voice to the silent language of culture: The *Cultura* project. *Language Learning & Technology*, *5*, 55–102.

George, A., & Sanders, M. (2017). Evaluating the potential of teacher-designed technology-based tasks for meaningful learning: Identifying needs for professional development. *Education and Information Technologies*, *6*(2), 2871–2895. doi:10.100710639-017-9609-y

Godwin-Jones, R. (2011). Mobile Apps for language learning. *Language Learning & Technology*, *15*(2), 2–11. Retrieved from http://llt.msu.edu/issues/june2011/emerging.pdf

Godwin-Jones, R. (2018). Chasing the butterfly effect: Informal language learning online as a complex system. *Language Learning & Technology*, *22*(2), 8–27.

Goodwin-Jones. (2005). Skype and podcasting: Disruptive technologies for language learning. *Language Learning & Technology*, *9*(3), 9–12.

Hannibal Jensen, S. (2019). Language learning in the wild: A young user perspective. *Language Learning & Technology*, *23*(1), 72–86.

Hyland, F. (2004). Learning autonomously: Contextualizing out-of-class English language learning. *Language Awareness*, *13*(3), 180–202. doi:10.1080/09658410408667094

Kennedy, I. G., Latham, G., & Jacinto, H. (Eds.). (2016). *Education skills for 21st century teachers: Voices from the global online educators' forum*. Cham: Springer. doi:10.1007/978-3-319-22608-8

Kessler, G. (2013). Collaborative language learning in co-constructed participatory culture. *CALICO Journal*, *30*(3), 307–322. doi:10.11139/cj.30.3.307-322

Lai, C. (2013). A framework for developing self-directed technology use for language learning. *Language Learning & Technology*, *17*(2), 100–122.

Lai, C. (2017). *Autonomous language learning with technology: Beyond the classroom*. New York, NY: Bloomsbury.

Lai, C., & Gu, M. (2011). Self-regulated out-of-class language learning with technology. *Computer Assisted Language Learning*, *24*(4), 317–335. doi:10.1080/09588221.2011.568417

Lai, C., Zhu, W., & Gong, G. (2015). Understanding the quality of out-of-class English learning. *TESOL Quarterly*, *49*(2), 278–308. doi:10.1002/tesq.171

Lee, J. S. (2018). Quantity and diversity of informal digital learning of English. *Language Learning & Technology*, *23*(1), 114–126.

Levy, M. (2009). Technologies in use for second language learning. *Modern Language Journal*, *93*(S1), 769–782. doi:10.1111/j.1540-4781.2009.00972.x

O'Dowd, R. (2018). From telecollaboration to virtual exchange: State-of-the-art and the role of UNICollaboration in moving forward. *Journal of Virtual Exchange*, *1*, 1–23. doi:10.14705/rpnet.2018.jve.1

OECD. (2015). *Students, computers and learning: Making the connection. PISA*. OECD Publishing.

Prensky, M. (2001). Digital Natives, Digital Immigrants. *On the Horizon*, *9*(5), 1–6. doi:10.1108/10748120110424816

Reinders, H., & Benson, P. (2017). Research agenda: Language learning beyond the classroom. *Language Teaching*, *50*(4), 561–578. doi:10.1017/S0261444817000192

Sauro, S., & Zourou, K. (2019). What are the digital wilds? *Language Learning & Technology*, *23*(1), 1–7.

Selwyn, N. (2013). *Education in a digital world: Global perspectives on technology and education*. New York: Routledge.

Sundqvist, P. (2009). *Extramural English matters: Out-of-school English and its impact on Swedish ninth graders' oral proficiency and vocabulary* (Unpublished Doctoral Dissertation). Faculty of Arts and Education (English), Karlstad University Studies, Karlstad, Sweden.

Sundqvist, P. (2015). About a boy: A gamer and L2 English speaker coming into being by use of self-access. *Studies in Self-Access Learning Journal*, *6*(4), 352–364.

Tar, P. M. (2017). *Exploring young learners' informal learning of English language: A comparative study on the perspectives of 11-13-year-old pupils in Finland and Vietnam* (Unpublished Master Dissertation). University of Eastern Finland, Finland.

Thomas, M. (2009). *Handbook of research on Web 2.0 and second language learning*. New York: IGI Reference. doi:10.4018/978-1-60566-190-2

Webb, M., & Cox, M. (2004). A review of pedagogy related to information and communications technology. *Technology, Pedagogy and Education*, *13*(3), 235–286. doi:10.1080/14759390400200183

Zhao, H. (2012). *World class learners: Educating creative and entrepreneurial students*. Sage Publications.

Section 1

Chapter 1
Informal language Practices via Digital Technology:
The Neglected Proficiency

Abdu M.Talib Al-kadi
https://orcid.org/0000-0003-3805-7507
Ibb University, Yemen

ABSTRACT

The main contention of this chapter is to dig into ICT-based serendipitous activities that second language (L2) learners perform beyond formal curriculum. It is based on an idea that juxtaposition of formal and informal language learning, under the auspices of ICT, broadens the ecology of learning and thus contributes to learners' overall proficiency. Nevertheless, formal language learning continues to be disconnected from practices that take place outside the classroom in hyperspaces, and the language uptake obtained from informal electronic involvement generally goes unnoticed. The chapter undertakes this missing proficiency and suggests implications to bridge or at least narrow the gap between formal and informal learning. It familiarizes teachers, parents, and course designers with today's learners' experiences of learning that occur after structured lessons. It implies that informal ICT-enabled practices should be fostered as supplementary and complementary to the formal instruction.

INTRODUCTION

Digital technology, which has become mainstream in modern life, has enlivened foreign language learning, making it cross barriers of the classroom and intertwine with informal learning. It has given language learners various and intriguing ways

DOI: 10.4018/978-1-7998-2116-8.ch001

Copyright © 2020, IGI Global. Copying or distributing in print or electronic forms without written permission of IGI Global is prohibited.

to acquire the target language in ways never attained before the emergence of sophisticated gadgets such as laptops, tablets, mobile phones, iPod, and an array of Web-based tools and applications. A large percentage of today's learners spend most of their lives connected to such electronic devices (Godwin-Jones, 2019; Ibrahim, 2018; Jarvis, 2014). The current generation of technology users, or as Jarvis dubbed the 'digital residents,' seem to digitize most aspects of their life, including second language (L2) learning. Since students' lives are permeated by a wide range of timely resources of information and communication technology (ICT), these 'digital residents' spend part of their lives performing ICT-based activities (Jarvis, 2014). They reside in cyberspace beyond formal learning contexts. Technology has saturated their lives and become a part of who and what they are (Godwin-Jones, 2018; Lee & Kim, 2014). Travis and Joseph (2009) noted that ICT empowers language learners to partake in virtual communication outside of class time; it has led to a diminution in their participation in formal education (Godwin-Jones, 2019). Arguably, the availability and diversity of such ICTs individualize practices – allowing users to manage learning activities by choosing the materials, time, and places of their study (Hegelheimer & O'Bryan, 2009). That is, the digital technology breakthroughs have spawned new opportunities of self-directing learning; learners self-monitor their learning, independently from the confinements of formal teaching (Bonk & Lee, 2017; Godwin-Jones, 2019; Lai, 2017).

Nevertheless, some researchers contend that the availability of ICTs does not mean that learners always embrace these learning opportunities. Carrying hand-held devices does not ensure that each of these technologies is used to the best effect (Levy, 2009). For instance, YouTube and Facebook which have matured into common sources for educational content are still a topic of controversy. Jarvis (2014) found that the participants in his study used ICT not to practice English explicitly but to perform free activities such as posting on Facebook, commenting on other users' posting or status and watching videos on YouTube. Similarly, Isbell (2018) found that there is little recognition of language learning activities in online spaces where language is the primary focus. The author argued that surfing the Internet for a general learning purpose does not automatically amount to what Krashen (2009) termed 'comprehensible input'. In the same vein, Lee (2019) alleged that L2 vocabulary development does not stem from simply placing learning in a technology-rich learning environment. It is in this spirit of query, the chapter considers the wider 'ecology' of learning, looking beyond what formal education offers. It hones in on the content of L2 learners' indulgence in electronic consumption and the relevance of this engagement to their English performance.

Rationale

Being a global contact language, English is often mediated through digital devices. English learners work on computers and mobile devices without the presence of teachers, or even without teachers' acquaintance with the electronic involvement that their learners are involved in. In some cases, ICTs may be the only medium of contact beyond the classroom (Jarvis, 2014). In the literature, the frequency of student-initiated ICT-based English activities is under researched (Ibrahim, 2013). Available studies on this theme reported that using technology in the classroom incentivizes students to continue learning beyond the teaching limited time and defined settings (Garrett, 2009; Isbell, 2018; Pereira, Fillol & Moura, 2019). Gonzalez and Louis (2013) provided an impulse to explicate how learners perform activities after school and how these activities contribute to their overall academic achievements.

The idea of this chapter stemmed from a line of prior research that called for informal learning (Alm, 2015; Al-Kadi, 2018; Brebera & Hlouskova, 2012; Chen & Bryer, 2012; Christine & Cathy, 2016; Lee, 2019; Trinder, 2017; Wang & Heffernan, 2009). These studies are based on a seemingly simple premise that digital ICTs allow L2 learners a wide variety of individual initiative that supports formal instruction. This work is also in congruence with Sharples, Arnedillo-Sanchez, Milrad, Vavoula's (2009) view that ICTs facilitate the interaction that people endure "with their surroundings, peers, and technology to create impromptu sites of learning and to carry their conversations from place to place, from time to time, from topic to topic" (p. 233). This bountiful research generally shows that students, when using digital technology, make general uses of it. Such studies simply scratched the surface of the phenomenon by touching on general uses of ICTs. The majority of prior research did not delve into measuring the informal learning outcomes. Using ICTs in sophisticated ways that pedagogues may value is still limited to a small percentage of learners (Chun, Kern & Smith, 2016; Godwin-Jones, 2018; Jarvis, 2014; Rogers, 2008; Sefton-Green, 2004; Trinder, 2017). The benefits, as well as challenges of appropriating ICT in everyday use for language learning, have not been well-explored, leaving wider room for research to explore uncharted areas. Informal technology-mediated exposure to English, which mostly goes unnoticed, is an ongoing heated debate. The putative value of informal learning must be weighed and assessed.

This is a point of departure to (a) dig into whether L2 learners have laudable informal ICT-based chances for L2 exposure beyond formally-structured learning and (b) elucidate the relevance of this informal learning to their overall performance. The chapter undertakes the issue of informal English practices through proliferate uses of modern digital tools and applications in a range of informal settings. For a better understanding, this endeavor deconstructs the topic into the following questions: In what ways does informal language learning affect formal language learning? What language skills/aspects do the learners of English elevate by informal consumption of digital technology? Is this electronic uptake taken into account when evaluating learners' overall language abilities? Are educators and proficiency testers aware of this effect? Does this informal language input go unchallenged? Do the existing language-ability tests measure the real proficiencies of L2 learners? With these questions in mind, the aim of this chapter is threefold. First, it encapsulates ICTs that learners frequently access. Second, it explores the extent this informal use of digital technology promotes formal English learning. Third, it ascertains whether English intake obtained through informal ICT use is wasted proficiency or is an added value to the existing formal learning opportunities.

Significance

The chapter is geared towards teachers, course designers, software developers, and linguists who look for ways of beefing English up and refining the perennial measurements of students' performance and proficiency. It genuinely informs this constituency of the target audience about today's learners' experiences of language learning that transcend the classroom. The chapter highlights how informal ICT uses complement and reinforce formal instruction. It affords L2 students useful ways to learn autonomously – which is one of the main, if not the prime, tenets of modern learning. Being familiar with learners' preferred ICT-based learning, teachers can tap into such potentials to help their learners make informed choices. By adopting the informal learning paradigm, technology-using teachers can decrease their workload and enrich their students' proficiency. Similarly, learning without overexerting or feeling bored, as it may happen in formal settings, students can enjoy learning that pleases them around the clock. Besides raising learners and teachers' awareness of the linguistic and pedagogical gains that learners obtain beyond the formal curriculum through free uses of electronic means, the chapter provides insights for devising measurements that take into account students' ICT-enabled English proficiencies. Additionally, the dearth of research in this area makes this work a significant contribution to bridge, or at least narrow, the schism between planned (formal) and non-intentional (informal) learning. It solidifies the background of the informal language learning through ICTs.

DIGITAL TECHNOLOGY: FROM CONSUMPTION TO PARTICIPATION

Information and communication technology (ICT), also known as digital technology, has become a lifeline for many. In this chapter, both terms are used interchangeably to treat technology as a vehicle for learning English. In this sense, it stands for well-connected devices, applications, and tools that serve a purpose in learning a foreign language. To mention a few, Web 2.0 tools such as Twitter, Facebook, wikis, blogs, and YouTube enable online participation in affinity spaces where users enjoy watching videos, listening to music, engaging in online learning communities (Godwin-Jones, 2018). An example of the most popular digital technology (the Web) has undergone three common phases: Web 1.0, Web 2.0, and the nascent Web 3.0. While Web 1.0 denied access to comment, respond, or quote, the second generation (Web 2.0) with its inbuilt features empowered social interaction, participation, and knowledge construction and sharing (MacLean & Elwood, 2009; Rüschoff, 2009; Wang & Heffernan, 2009). Geared around socio-cultural exposition, worldwide interaction, and knowledge sharing, Web 2.0 is "essentially a transition from the 'online consumer' to the 'participant' producer" (MacLean & Elwood, 2009, p.475). Being user-led and community-based with collaborative content (Wang & Heffernan, 2009), Web 2.0 is viewed as a unique platform for language learning and teaching. The extension of the Web has created new applications for a wider range of participatory roles; YouTube, blogs, and wikis enable users to collaborate, criticize, and comment on electronic content. At the time this work was underway, the growing Web 3.0 and Web 4.0 are expected to advance the process of interaction and participation to unimaginable extents. It is rather a move towards making content accessible by multiple non-browser applications, the thrust of artificial intelligence technologies, the semantic web, the Geospatial Web, or Web 3D. The crux of the matter within these web generations is that students, working on any device, learn or practice the target language on their own at any time.

With the participatory roles of the new version of the Web, learners become producers rather than consumers of knowledge (MacLean & Elwood, 2009). Educators would benefit richly from this feature by shifting a focus on language learners' everyday use of the web-based applications inside and outside of classrooms. These tools/Apps are accessed through a plethora of wearable gadgets and media (TV, radio, web-based channels). Although these embed sorts of informal learning activities, this phenomenon is inadequately versed; informal language learning continues to be underrepresented in the body of research. The fusion of digital technologies into formal and informal ICT-based learning, in Christine and Cathy's (2016) words, is still "under-theorized or treated as binary conditions, which oversimplify the complexities of the actual learning contexts today's youth inhabit" (p.7).

FORMAL VS. INFORMAL LEARNING

Before 'electronic learning', informal learning has been conceived as untrue learning. The rise of e-learning gave birth to a wide range of informal activities, and informal learning has been wrapped into the general notion of e-learning since then. There is now an increasing reliance on informal learning as it complements formal learning. They are different shades of the same phenomenon. Nonetheless, an authoritative definition of informal learning barely exists and it is always defined in contrast to formal learning in several dimensions. It has long been assumed that formal and informal learning are distinct in some ways. For instance, formal learning takes place within an organized (structured) context with credit courses (Eaton, 2011). Following Eaton's definition, formal learning is a process of instructor-led education wherein activities are purposefully assigned by formal representatives (e.g. teachers) who prepare and guide ICT-based lectures, activities, language lab sessions, seminars – which are all essentially structured within specific time and place. On the contrary, informal learning occurs beyond the classroom routine. It is loosely described as impromptu learning – disorganized, experiential, and unguided learning. Seen in another light, it is unprompted learning that lacks intention from the learners' standpoint (Eaton, 2011). Sockett (2013) described informal language learning as 'generally incidental' with a focus on "activities being communication and enjoyment rather than language learning" (p. 49). But when learners study without teachers, they are assumed to be autonomous learners. This is a reason to ask: are all learners autonomous?

In recent research, informal learning is inseparable from the overall learning, rather than contrasting it with formal learning. Rogers (2014) viewed it as "lying on a continuum ranging from accidental/incidental learning, through task-conscious learning, through self-directed learning to non-formal and formal learning" (p.10). Interestingly, Rogers likened formal and informal learning to an iceberg – the tip of it represents the conscious formal learning; and the biggest part of it, which is under the water, stands for informal learning (unseen). Drawing on Rogers' simile, learning is boundless to formal settings (e.g. schools and universities). It happens in an extended learning environment through day-to-day activities. If we accept Rogers' (2014) image of the iceberg, informal learning which is bigger than formal learning remains unmeasured. That is to say, a considerable portion of learners' proficiency remains untested – it is missing proficiency. Thereby, there is an inconsistency between what learners actually practice and what is actually measured by formal assessments. Formal tests measure only the top of the iceberg. In McNamara's (2008) words, "a proficiency test…is an assessment of what has been learned of a known or an unknown syllabus" (p. 88). If the 'known' syllabus is a formal curriculum, the latter is informal learning. While the formal paradigm is based on methodology, informal learning draws on theories of personal and self-regulated approaches. It may take place anywhere with no time constraints. The mobility of learners and learning has complicated learning as well as testing. This complexity, besides lack of qualified and competent assessors of informal language uptake, continues to be a major challenge.

Another point of contention is related to informal learning itself. Learning that takes place beyond class hours has been traditionally categorized as informal, non-formal, incidental, autonomous, lifelong learning (Lai, 2017). All these forms of informality lack a proper definition as there is a flawed understanding of the term. Sefton-Green (2004) raised questions whether informal learning simply means learning that happens in different ways in different places, or it is anything learnt yet educational systems devalue? It is nebulous whether 'informality' refers to methods of learning, places of learning, the content of learning, or the relationship between the activity and what is valued as knowledge. Whatever the difference, informal learning, though unruly, remains the most extensive learning resulting from common activities (Brebera & Hlouskova, 2012). If what Latchem (2014) says is true, 70 to 90% of human learning is determined individually. This indicates a large power afforded to informal learning.

Although it is being rapidly streamlined with formal education in the worldwide context, there is little serious attention of how ICT-based language learning is foundational for L2 formal learning (Blake, 2008; Latchem, 2014; Sockett, 2014; Pereira, Fillol, & Moura, 2019). Previous studies on ICT-based formal and informal learning have come up with inconclusive findings. A hiatus currently exists between the two learning modes, or let us say between the 'controlled' and 'less controlled' learning. Perhaps, this is because embracing opportunities that L2 students reside to express themselves in an ICT-based environment is a thorny issue. The burgeoning of ICT provided an incentive for exploring uncharted territories.

Figure 1. Formal and informal learning iceberg

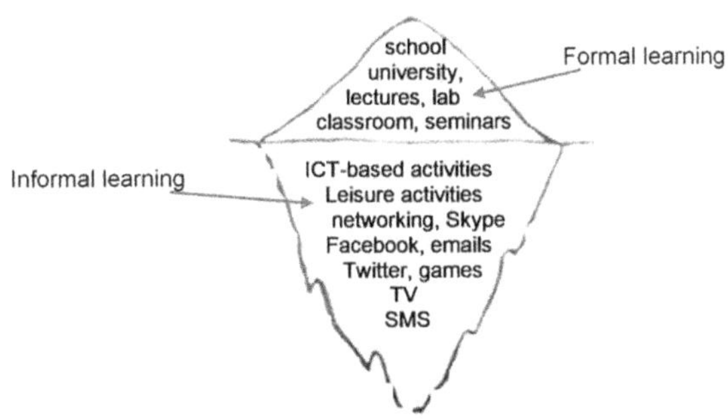

Fusion of Formal and Informal Learning

The waves of digital technology are increasingly credited as mediums of learning which have made the line between formal and informal language learning rather blurred (Christine & Cathy, 2016; Rogers, 2014). Informal learning is no more seen as an inferior form of learning, and it is widely acknowledged as supplementary to formal education. It mends and complements formal learning. It makes for possible shortcomings of formal language learning in terms of exposure and authenticity (Godwin-Jones, 2018; Jarvis, 2014; Meskill & Quah, 2013). Actually, both learning modes are overlapped, and it is difficult to justify their divorce from each other. A big deal of formal learning today involves elements of informality. Equally, in some informal learning situations, there are some elements of formality, i.e. formalizing informal learning and informalizing formal learning. Informal learning may be explicit and implicit; the former differs from its counterpart by learner's awareness of the learning outcomes.

The stance adopted in this chapter favors formalizing informal learning within the influx of digital technology. Technology-enhanced language learning (TELL) has been diffused across the world under several labels: online learning, electronic learning, and technology-based learning. Formal ICT-enabled learning is defined here as a situation wherein EFL teachers prepare teacher-oriented classes that include activities deliberately designed and assigned to EFL learners in scheduled classes. This might straddle classroom lecturing, classroom discussion, seminars, and language lab sessions in which using ICT is patently evident. Nonetheless, there is no shortage of criticism of educational systems that shy away from using technology for learning purposes. Tridner (2017) contends that technology in formal classrooms "does not represent the most effective use of contact time" (p. 410). What exacerbates the matter is that the role of ICT in formal settings is foggy as it lacks scope and sequence (Al-Kadi, 2018; Chapelle, 2007). Even accessible ICTs inside the classroom, in most cases, are handled unsystematically. Teachers may use technology mechanically to serve a short-scale purpose and enhance certain aspects of the language. Hence, linguistic abilities are bound to be limited. Outside the classroom, today's learners "engage with technologies in ways that are often more varied and more sophisticated than those they encounter at school" (Chun, Kern & Smith, 2016, p.76). Arguably, ICT-based serendipitous activities increase opportunities of L2 uptake although ICT is not considered to be educational – from the conventional understanding of it (Sefton-Green, 2004). The normalization of ICT applications gives rise to the question of whether informal learning is still arbitrary and pointless. Sefton-Green noted that informal learning now does not suggest that informal approaches to learning are all fun and games, and 'formal' approaches are all seriousness and gravity. What really matters is the purpose and configuration of learning experience in both learning modes.

CLOSING THE SCHISM

Jarvis surmised that "with English dominating the Internet, with ever-growing numbers of connected learners and users, and with a long tradition of innovation, TESOL is uniquely placed to rise to such challenges" (Jarvis, 2014, p. 31). There have been numerous attempts in the worldwide context to live up to the challenge. To begin with, Ibrahim (2013) examined (a) the trends and patterns of ESL learners' engagement with the Web 2.0 tools and (b) the perceived affordances and limitations to the Web-based practices in the Malaysian context. The results indicated positive engagement and robust endorsement of the Web for supporting ESL learning besides enhancing social interaction. For instance, repetitive practice, fun, authentic and effortless learning, and positive feedback were found supportive for English learning. Likewise, Wang (2014) studied the possibility of linking formal teaching with informal practices employing web-based role-playing simulations. The results accounted for EFL students' engagement in an authentic social interplay that provides laudable benefits for promoting their communicative abilities and interest in foreign language learning. In the same line of research, Jarvis (2017) explored through a questionnaire and focus group non-native speakers' (NNs) perceptions and informal practices of digital technology in the Thai context. The study emphasized the ubiquity of mobile-assisted language use to familiarize students with new challenges facing the ELT profession such as plagiarism and digital footprints.

Chances of informal learning are proliferating online. In contexts where exposure to the target language is limited, learners may opt for learning online through a wave of ventures such as Khan Academy, youGlish, FutureLearn, Coursera, Udemy, to mention just a few. ICT-based learning projects are now handy. For instance, Travis and Joseph (2009) developed a podcast-based project to elevate the speaking skills of advanced learners. The project was based on a premise that web-based tools (podcast as a case in point) give way for user-generated content and mass participation. The results showed that students who used the Internet selectively with high skill level aspired remarkably. Similarly, Hegelheimer and O'Bryan (2009) examined podcasts meant for self-study with various levels: ranging from native language speakers (without formal teacher training) to university-level language educators. In virtual spaces, informal learning involves participation in social networking sites as well. For instance, Sockett (2011) examined online English learning in the French context. Over twelve months, the study explored major activities that French EFL learners performed informally, and it showed an abundance of informal listening activities (a corpus of television series) that the informants performed in their leisure time. It also showed the relevance of this input material for formal learning. Likewise, Alm (2015) examined how 190 New Zealander language students at the university level used Facebook, viewing it as a multilingual environment. The results indicated that students used a range of Facebook features for communication with native-speaker Facebook friends. Taken together, this line of research shows that informal learning is possibly linked with formal teaching through an array of ICT tools and applications, as illustrated in Figure 2. Reporting from the Czech Republic, Brebera and Hlouskova (2012) provided suggestions that can be useful to bridge the gap including internet forums, portfolio assignments, and online games. These models may be used to theorize the web as a space for informal learning.

For the most part, it is reasonable to caution against overly optimistic expectations from informal uses of technology. For one thing, free and uncontrolled ICT informal uses "involve the messy, unpredictable use of human language for motivated, authentic purposes, a phenomenon that does not lend itself to laboratory controls" (Meskill & Quah, 2013, p. 41). This encouraged authors to explore the association between free use of ICTs and users' linguistic achievements. Lee (2019) examined the correlation between informal digital English learning and vocabulary acquisition in the Korean EFL context; recruiting 77 university students of different majors and employing a questionnaire and interview, the study concluded that a digital-rich learning situation does not guarantee an automatic successful development of L2 vocabulary – a view endorsed by Isbell (2018) who asserted that surfing the Internet for a general learning purpose does not automatically turn into fruitful language learning. For another thing, ICT has brought about unusual linguistic norms – so long as English is concerned– such as shorthand typing, abbreviations, emoticons, and instant message lingo (Al-Kadi, 2018; Crystal, 2006). This has created a reaction that the growing weird variety of English may weaken the standard form of English language. The evolving internetese, emailese, and textese are all examples of this new unusual usage of language.

Figure 2. Connecting formal and informal language learning through technology

Language Learning and Proficiency

Touched on above, the robust evidence of noteworthy outcomes of informal language learning brings to the core the concept of second language learning and proficiency. Language learning that takes place primarily in formal settings is contrasted to incidental learning that happens in real interactions (Bahrani & Tam, 2012; Krashen, 1985, 2009). Believingly, the greater the magnitude of exposure, the quicker the acquisition (Blake, 2008; Krashen, 2015). Dedicated practice is necessary for achieving a high level of proficiency. This dedication stems from not only formal instruction but also informal learning opportunities. That is, in order to supplement classroom learning, continued, self-initiated informal learning is deemed necessary. Chances are that learners acquire language informally via free and timely access to ICT tools and applications (Alm, 2015; Chen, 2013; Godwin-Jones, 2018; Jarvis, 2014; Trinder, 2017).

Arguably, learners who incorporate ICT in their English learning outperform those who do not go for digital technology incorporation (BenYoussef & Dahmani, 2008). However, it is hard to imagine that there are learners now who are unacquainted with modern technologies. Rather, it is reasonable to assume that some individuals are more digitally-literate (tech-savvy) than some others. Whatever the case, it is generally accepted that English bounds to evolve and flourish wherever ICT is used, and this presupposes progress in English learning (Blake, 2008; Garrett, 2009; Wang, 2014). Trinder (2017) noted that "casual interactions revolve around input obtained from mobile devices" (p. 402) as these hand-held gizmos provide effective virtual spaces for international communication and learning. The author provided insights into this landscape of research arguing that such technologies boost informal language learning, raise awareness about the utility of underused resources, and foster strategies for exploiting ICTs. In addition, Rüschoff's (2009) findings provided more impetus for exploration language output through digital platforms. The author indicated that students express themselves informally on Facebook, Twitter, and other social media.

These claims are broadly guided by Krashen's (2009, 2015) input hypothesis and Swain's (1985) output hypothesis which have dominated the field of L2 learning and acquisition for years. Nonetheless, educators may undervalue this free and unstructured learning (Pereira, Fillol & Moura, 2019) which is gaining a worldwide acceptance as self-motivating, self-monitored, self-paced, and self-assessed learning opportunities that institutions of formal learning hardly accommodate satisfactorily (Costa, Cuzzocrea & Nuzzaci, 2014; Godwin-Jones, 2019; Ibrahim, 2013; Isbell, 2018; Olek-Taszarek, 2014). By using ICTs beyond the classroom, the language learners are immersed in English of different varieties that formal educational institutions fail to provide, particularly in contexts where L2 learners have insufficient time of instruction. That is to say, formal education generally fails to furnish learners with sufficient comprehensible input of the target language. Formal teaching per se is not enough to achieve high levels of proficiency.

In fact, the distinction between language learning and acquisition in the era of modern ICTs is rather fuzzy. Guth (2009) maintained that "language acquisition…is a life-long process that cannot end with traditional education and must be cultivated throughout life often through informal learning in different contexts and situations" (p. 453). Language learning, whether it be instructed or autonomous, "remains in spite of all curricular plans/logical syllabi nonlinear and unpredictable" (Godwin-Jones, 2018, p. 20). The author supported the emergent outcome, which is often tacit, incidental, and amorphous. This unpredictable learning fits in with the ultimate goal of technology-integration: encouraging spontaneous learning. The traditional classroom setting is only one of several venues of language acquisition. It may be facilitated by interplay with native speakers or through exposure to authentic ICT-mediated input (Bahrani & Tam, 2012). Suffice it to say that using ICTs outside the classroom is tension-free, pleasurable, and self-managed. In Godwin-Jones' (2018) words,

Viewing students as engaged and authentic members of society- not just as L2 learners- makes the language experience potentially more meaningful to their personal lives and may lead them to see their L2 as an integral part of who they are. (p. 22)

This way, learners' lives and experiences become the cornerstone of their learning. Huang (2008) found that students who scored better on standardized achievement tests were those who used computers more often at home and less at school, i.e. home use, not school use, was associated with greater achievement. In this regard, Caffesse and Guasch (2010) argued that healthy utilization of ICT provides a learning platform where EFL becomes ESL in terms of exposure to the target language (TL). According to the Caffesse and Guasch, English under the influence of ICT in many countries has become a second language rather than a foreign language which is a good source of language acquisition (Blake, 2008; Guth, 2009; Sockette, 2014).

Evaluation research seeks direct evidence of language learning resulting from ICT-based activities (Chapelle, 2007). With an increasing reliance on language learning through ICT, a pending question pops up: Do formal tests really measure the actual proficiencies of tech-savvy learners? For long, learners of English all over the world have been tested on a variety of English that they do not and, in some cases, never use. Griffith and Lim (2010) rightly stated that

Students are assessed based on how well they perform on written tests (usually grammar-based), and on the basis of these tests receive an evaluation of passing

or failing with the requisite awards or penalties. Teachers, in turn, are evaluated on the basis of how well their students perform on these tests and are labeled as good or bad teachers. (p.75)

Having said that, Griffith and Lim get into a quite telling argument that many learners take proficiency tests because administrators, teachers, and parents take it for granted that testing is useful for some job vacancies or grants which require proficiency tests of a certain type. However, changes in the status of English today and the emerging Englishes have made it difficult to decide which variety of English to test: English as a second language (ESL), English as a foreign language (EFL), English as a lingua franca (ELF), English as an international language (EIL), and there are other varieties in the pipeline.

The Decline of Native-Speakerism

While classifying English as EFL and ESL was governed by a colonial factor, the modern impulses to learn English, the need and importance of English have been re-defined (García, 2013; Mesthrie & Bhatt, 2008). The new labels ELF and EIL have been determined by the power of science and technology which is an important factor that has guided a modification in the linguistic map of modern English. The growing power of globalization, economy, communication, and culture has made EFL prominent in countries where English was uncommon such as China, Russia, and Brazil. English, under the influence of globalization, has gained a general context wherein English is a worldwide contact language. It is the language of international conferences where hundreds of non-native speakers of English who attend such gatherings contact through it. Hence, English – being the lingua franca of the world now– differs from EFL in some ways. While EFL requires meeting the native-speakerism norms, ELF hinges on globalized English. The former (i.e. EFL) gives eminence to the native speaker' cultural aspects, unlike ELF which has earned a worldwide context. That is, ELF is broader than its predecessors: ESL and EFL (García, 2013; Graddol, 2006). Language learners now have more opportunities to learn a culture and a language that technology has made global. From the SLA stance, this is good for the provision of high quality of English input. It is difficult to disassociate ICTs in formal contexts from informal technology uses. ICT-oriented activities are sources of daily informal language exposure which is inseparable from the overall English proficiency; students gain linguistic and pedagogical benefits from technology integration in their learning (Bahrani &Tam, 2012). However, measuring these benefits is still a topic of debate. Gauging informal learning outcomes separately from formal tests is rather impractical. Equally, it is difficult to measure informal learning via formal tests.

What emerges from the foregoing discussion is that the existing proficiency tests generally measure proficiency that meets the native-speaker's model and tend to neglect linguistic gains accruing from informal learning situations. Many English proficiency-test boards still use the native norms in designing proficiency tests. In fact, the validity of tests that meet the native modes has been weakened. García (2013) and Mesthrie and Bhatt (2008) argued that the dominance of 'native-like' proficiency has lost ground in the worldwide context. Godwin-Jones (2018) provided reasons to refrain "from putting the native speaker forward as a model and goal of language instruction" (p. 21). For one thing, the linguistic map of the English language has changed under the influence of modern technology, and thus tests should change too. Proficiency test takers are being tested on British English or American English, and not on the international English that they actually use. Apart from the world Englishes, emerging varieties of English include emailese, internetese and textese that Crystal (2013) believe are additions to the standard variety. All of these varieties of modern English are what is argued – in this chapter – as unexploited proficiency. One more thing, the non-native speakers (NNSs) and readers of English today are a majority compared to those who use it as their L1 (Chauhan, Ying & Zhenfang, 2013). Let alone, the worldwide interaction occurs in the absence of native speakers (Graddol, 2006). By virtue of this incontestable fact, Crystal (2006) conceded that "no one nation can any longer be said to "own English" (p. 412). It is everybody's language. In the same vein, Jarvis (2014) noted that the bulk of 'digital residents' in the international arena are NNSs of English who, as a consequence of English being the lingua franca of the World, live out "a part of their lives in both their first language (L1) and English as their L2" (p. 22). Again, English today is widely used by "multilingual speakers whose first language is other than English" (García, 2013, p.114). They use a myriad of electronic gadgets on a daily basis: mobile phones, tablets, iPod, etc. Using such hand-held-devices has morphed into a daily routine. They chat on messenger, post on Facebook, twit on Twitter, and get in touch via Skype, WhatsApp, and the like. Arguably, these digital gadgets give learners access to a virtual world to use the target language in sharing interests and hobbies.

Be it formal or informal, a considerable English language use nowadays is technology-mediated (Jarvis, 2014; Trinder, 2017), primarily outside the realm of formal education. Even so, this linguistic gain remains ignored and not included in formal tests. Formal tests generally measure the formal English that learners have to acquire for formal certificates. Jenkins (2006), in her article about the role of EIL to the practice of English language proficiency test, underlined the fact that the use of English outside the inner circle of English users has now become too massive to be ignored in designing English proficiency tests. In fact, measuring proficiency in midst of diversified needs and varieties of English in the global arena is a sophisticated mission. Due to changes in the status of English and expanding the spectrum of learning ecology and CALL, the question arises: Which variety of English is to be tested in order to determine the proficiency of learners? ESL, EFL, ELF, EIL or all of these?

Since most of the language learning today comes from informal practices of the target language in the virtual world (through digital technology), there seems to be no compelling reason to accept measuring proficiency on the norms of the native model and the formal variety per se. The passable access to sophisticated devices and powerful tools promotes language instruction (Sharples et al., 2009) and makes possible to host various learning opportunities (Sefton-Green, 2004). This subsumes learning that takes place outside formal institutions as part of leisure activity rather than being exam-oriented. In Blake's (2014) words, "today's FL curriculum encompasses not only the time spent in class…but also the effort spent outside the classroom working in groups (with or without contact with the target speech community at large), as well as all those moments of the night and day spent alone, quietly studying the target language (p. 131). All these moments and avenues are imbued with technological innovations. Thanks to digital technologies, language activities are now technology-mediated, and this integration has become the linchpin of students' social and cultural lives; yet such activities are "viewed by formal educational establishments as outside the realm of educational experience" (p. 8). Thus, measuring learners' proficiency without the above-mentioned considerations provides inaccurate pointers of learners' real proficiency, and for a better understating of it, the learning moments during which learners employ technology on their own should be included (Ben Youssef & Dahmani, 2008; Lai, 2017).

IMPLICATIONS

This chapter breathes life into some implications for researchers, pedagogues, and decision-makers to mitigate the gap between formal and informal learning. These implications can make a paradigm shift as illustrated in Figure 3.

From the Classroom to Learning Spaces

An important implication this chapter brings to the core is enlarging learning ecology. All accessible ICTs create an unregulated learning environment which is boundless to classrooms. The more the advent of ICT appliances, the more the informal learning get expanded. The term 'learning spaces' supersedes the traditional learning venue called 'classroom' (Jahnke, et al. 2012); the former seems to be more appropriate to host all the emergent learning opportunities. This extended ecology of learning comprises schools, homes, playtime, libraries, museums, and the virtual world (online presence) – all of which play their role in today's learning (Godwin-Jones, 2018; Reinders & Benson, 2017; Sefton-Green, 2004). The classroom learning is only one variable within a larger endeavor (Garrett, 2009; Godwin-Jones, 2018; Sharples et al., 2009). Besides pair work and group work, learning spaces are good for networking – an environment where learners are exposed to different varieties of English, international users of English and a wide range of language skills and activities.

From CALL to ALL

Broadening the learning environment necessitates extending the scope of the traditional concept of technology-based learning. The practices and perceptions accruing from prior research suggest a shift from CALL and its associated allies MALL, WELL, and IALL to a new paradigm that hosts all the structured and unstructured technology-enhanced practices, including all possibilities of learning from, thorough, and around technology (formally and informally). Taking "less-controlled or informal learning contexts and the data on how students use a range of devices, for what purposes, and the role of English as an L2 in such practices" (Jarvis, 2014, p. 28), CALL by itself does not represent the whole picture of TELL. By the virtue of extended learning ecology, including informal opportunities of learning L2, the acronym ALL which stands for "All Language Learning" (Jarvis, 2014, p. 28) nicely appeals to the challenge.

New Measurements of Proficiency

Another point this chapter highlights is that a lack of adequate tests to measure a broader range of learners' communicative abilities and remains a desire. Informal English learning through technology is neglected and thus becomes missing proficiency. This endeavor implies that there should be new ways to assess learners' ICT-based activities, which have become an essential part of language learning. It

Figure 3. Shifting from formal to informal language learning through digital technology

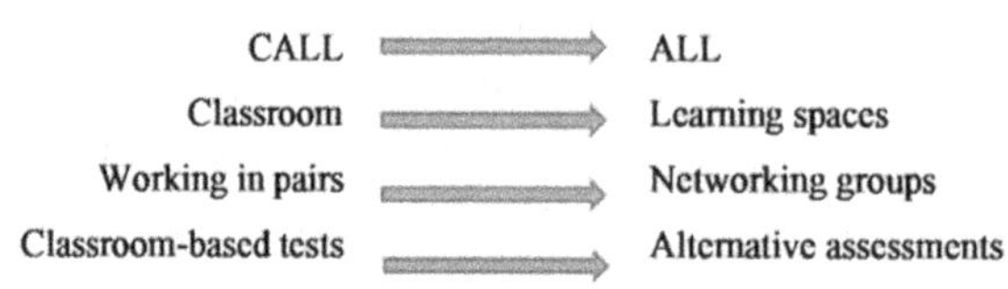

is now difficult to determine one's proficiency by ordinary tests. The tests at hand (e.g. IELTS, TOEFL) might not gauge exactly the actual proficiency of English learners. In other words, proficiency should be tested from a wider perspective rather than measuring proficiency for meeting native speaker's notion. It is quite uncontroversial that most language learning occurs beyond formal teaching. The profuse flow of ICTs has fueled this change, making informal learning overlap with formal education (Christine & Cathy, 2016; Rogers, 2014). To reiterate, with students nowadays connected to a variety of social media, online participation in affinity spaces (watching videos, listening to music, engaging in online learning communities), it is unreasonable to test only English that learners are exposed to in formal settings – that meets the native speakers' model. Standard tests generally measure formal and standard language but not the varieties acquired informally. Proficiency should be measured on the basis of both self-assessed and formal grading criteria.

Taken together, these implications call for recognizing language learning that takes place beyond the formal curriculum (extracurricular resource) as an integral part of L2 learners' overall performance. It is a stepping-stone for shifting from teacher-led to student-initiated learning. This shift is not always straightforward and cannot happen overnight. Relating student-initiated English practices with formal English learning objectives requires new pedagogies and delicate analyses. There should be a gradual transition toward it. There are several ways for teachers and learners to expand the learning environment and link formal and informal learning. For instance, learners may use ICTs freely to debate homework tasks or share ideas among their peers and/or teachers. Teachers may set homework tasks to be handled via some sort of ICTs in such a way to (a) promote and develop connections with home uses of technology, and (b) enable students to identify suitable resources. Hegelheimer and O'Bryan (2009) recommended that "any out-of-class material to be truly integrative, it must serve a course-related function, whether reviewing lecture content, elaborating upon difficult concepts, or preparing students for the next class" (p. 342).

CONCLUSION

Bringing prior research together, the chapter marshals evidence of optimizing English learning through informal proliferate uses of ICT devices, tools, and applications. It accumulated evidence on L2 development outside institutional milieu, predominately with an array of digital technologies. This opens a wide gate for learning independently from formal authorities. This informal intake benefits learners in enhancing their overall proficiency. It implicitly affects learners' performance and helps them perform well in formal tests, not always though. They develop oral skills and broaden their cultural awareness of the target language. Besides insights into incorporating informal ICT-based practices as potent resources in formal situations within an inclusive pedagogy and egging on students to utilize it fruitfully, the chapter elucidated how informal language intake enriches the overall performance of learners of English. Insights on learners' experiences, motivation, and interests in self-directed uses of ICTs explain how technology helps learners optimize linguistics aspects, acculturation, and other related issues. Nevertheless, informal language uptake requires formal recognition. Otherwise, it becomes unexploited expertise. Despite the fact that TELL constitutes a big deal of modern learning habits, chances of informal learning are not meant to obviate formal instruction. Neither should it be viewed as an elixir for all formal learning inadequacies. Rather, both learning modes complement each other in an inspiring and motivating ways.

REFERENCES

Al-Kadi, A. (2018). *Some Aspects of ICT Uses in the Teaching of EFL at the Tertiary Level in Yemen* (PhD dissertation). ISLT, Carthage University, Tunisia.

Alm, A. (2015). Facebook for informal language learning: Perspectives from tertiary language students. *The EUROCALL Review*, *23*(2), 3–18. doi:10.4995/eurocall.2015.4665

Bahrani, T., & Tam, S. (2012). Informal language learning setting: Technology or social interaction? *The Turkish Online Journal of Educational Technology*, *11*(2), 142–149.

Ben Youssef, A., & Dahmani, M. (2008). The impact of ICT on student performance in higher Education: Direct effects, indirect effects and organizational change. *RUSC: Universities and Knowledge Society Journal, Fundació Universitat Oberta de Catalunya,* *5*(1), 45-56. https://halshs.archives-ouvertes.fr/halshs-00936560

Blake, R. (2008). *Brave new digital classroom: Technology and foreign language learning*. Washington, DC: Georgetown University Press.

Blake, R. (2016). Technology and the four skills. *Language Learning & Technology*, *20*(2), 129–142. Retrieved from http://www.lltjournal.org/item/2951

Bonk, C., & Lee, M. (2017). Motivations, achievements, and challenges of self-directed informal learners in open educational environments and MOOCs. *Journal of Learning for Development*, *4*(1), 36–57.

Brebera, P., & Hlouskova, J. (2012). Searching for bridges between formal and informal language education. *IADIS International Conference on Cognition and Exploratory Learning in Digital Age (CELDA)*, 274- 277.

Caffesse, L., & Guasch, T. (2010). Foreign languages turned into second languages through ICTs. Academic Press.

Chapelle, C. (2007). Second language learning and online communication. In R. Andrews & C. Haythornthwaite (Eds.), *The SAGE handbook of e-learning research* (pp. 371–393). SAGE Publications. doi:10.4135/9781848607859.n17

Chen, B., & Bryer, T. (2012). Investigating instructional strategies for using social media in formal and informal learning. *International Review of Research in Open and Distance Learning*, *13*(1), 87–104. doi:10.19173/irrodl.v13i1.1027

Chen, X. C. (2013). Tablets for informal language learning: Student usage and attitudes. *Language Learning & Technology*, *17*(1), 20–36.

Christine, G., & Cathy, L. (2016). Social media and education: Re-conceptualizing the boundaries of formal and informal learning. *Learning, Media and Technology*, *41*(1), 6–30. doi:10.1080/17439884.2015.1064954

Chun, D., Kern, R., & Smith, B. (2016). Technology in language use, language teaching, and language learning. *Modern Language Journal*, *100*(S1), 64–80. doi:10.1111/modl.12302

Costa, S., Cuzzocrea, F., & Nuzzaci, A. (2014). Uses of the Internet in educative informal contexts. Implication for formal education. *Media Education Research Journal*, *43*(22), 163–171. doi:10.3916/C43-2014-16

Crystal, D. (2006). *Language and the Internet* (2nd ed.). Cambridge, UK: CUP. doi:10.1017/CBO9780511487002

Crystal, D. (2013, November 9). *The effect of new technologies on English (from the interview with David Crystal in Belgrade)*. Available on https://www.youtube.com/watch?v=qVqcoB798Is

Eaton, S. (2010). *Formal, non-formal and informal learning: The case of literacy, essential skills, and language learning in Canada*. Retrieved from ERIC database. (ED508254)

Eaton, S. (2011). *How long does it take to learn a second language? Applying the "10,000-hour rule" as a model for fluency*. Retrieved from ERIC database. (ED516761)

García, R. (2013). English as an international language: A review of the literature. *Colombian Applied Linguistics Journal, 15*(1), 113–127. doi:10.14483/udistrital.jour.calj.2013.1.a08

Garrett, N. (2009). Technology in the service of language learning: Trends and issues. *Modern Language Journal, 93*, 697–718. doi:10.1111/j.1540-4781.2009.00968.x

Godwin-Jones, R. (2018). Chasing the butterfly effect: Informal language learning online as a complex system. *Language Learning & Technology*, *22*(2), 8–27.

Godwin-Jones, R. (2019). Riding the digital wilds: Learner autonomy and informal language learning. *Language Learning & Technology*, *23*(1), 8–25.

Gonzalez, D., & Louis, R. (2013). CALL in low-tech contexts. In M. Thomas, H. Reinders, & M. Warschauer (Eds.), *Contemporary computer-assisted language learning* (pp. 217–242). London: Bloomsbury Academic.

Graddol, D. (2006). *English next: Why global English may mean the end of English as a foreign language*. London: The British Council.

Griffith, W., & Lim, H. (2010). Making student-centered teaching work. *MEXTESOL Journal, 34*(1), 75-82. Retrieved from https://www.mextesol.net/journal/index.php?page=journal&id_article=37

Guth, S. (2009). Personal learning environments for language learning. In M. Thomas (Ed.), *Handbook of research on Web 2.0 and second language learning* (pp. 451–471). Hershey, PA: Information Science References. doi:10.4018/978-1-60566-190-2.ch024

Hegelheimer & O'Bryan. (2009). Mobile technologies, podcasting and language education. In M. Thomas (Ed.), *Handbook of research on Web 2.0 and second language learning* (p. 331). Hershey, PA: Information Science References. doi:10.4018/978-1-60566-190-2.ch018

Huang, S. H. (2008). *The relationship between computer use and academic achievements* (Ph.D. dissertation). The University of North Texas.

Ibrahim, K. H. (2018). A methodological approach to analyzing digital game-based FL use and learning: The diamond reconstruction model. In B. Zou & M. Thomas (Eds.), *Handbook of research on integrating technology into contemporary language learning and teaching* (pp. 405–426). Hershey, PA: IGI Global. doi:10.4018/978-1-5225-5140-9.ch020

Isbell, D. (2018). Online informal language learning: Insights from a Korean learning community. *Language Learning & Technology*, *22*(3), 82–102.

Jahnke, I., Bergström, P., Lindwall, K., Mårell-Olsson, E., Olsson, A., Paulsson, F., & Vinnervik, P. (2012). Understanding, reflecting and designing learning spaces of tomorrow. In I. A. Sánchez & P. Isaías (Eds.), *Proceedings of IADIS mobile learning* (pp. 147–156). Retrieved from http://www.mlearning-conf.org/

Jarvis, H. (2014). Digital residents: Practices and perceptions of non-native speakers. *The Asian EFL Journal, 75*, 21–35.

Jenkins, J. (2006). Current perspectives on teaching World Englishes and English as an international language. *TESOL Quarterly, 40*, 157–181. doi:10.2307/40264515

Krashen, S. (1985). *The input hypothesis: Issues and implications*. New York: Longman.

Krashen, S. (2009). The comprehension hypothesis extended. In T. Piske & M. Young-Scholten (Eds.), Input matters in SLA (pp. 81-94). Bristol, UK: Multilingual Matters.

Krashen, S. (2015, Feb. 17). *British Council interviews with Stephen Krashen (part 3 of 3)*. Available on YouTube: https://www.youtube.com/watch?v=EzBl6jBtf6E

Lai, C. (2017). *Autonomous language learning with technology beyond the classroom*. London: Bloomsbury Publishing Plc.

Latchem, C. R. (2014). Informal learning and non-formal education for development. *The Journal of Learning for Development, 1*(1). Retrieved from https://files.eric.ed.gov/fulltext/EJ1106082.pdf

Lee, J. (2019). Informal digital learning of English and second language vocabulary outcomes: Can quantity conquer quality? *British Journal of Educational Technology, 50*(2), 767–778. doi:10.1111/bjet.12599

Lee, J., & Kim, H. (2014). An exploratory study on the digital identity formation of Korean University EFL Learners. *English Teaching, 13*(3), 149–172.

Levy, M. (2009). Technologies in use for second language learning. *Modern Language Journal, 93*, 769–782. doi:10.1111/j.1540-4781.2009.00972.x

MacLean, G., & Elwood, J. (2009). Digital natives, learner perceptions and the use of ICT. In M. Thomas (Ed.), *Handbook of research on Web 2.0 and second language learning* (pp. 156–179). Hershey, PA: Information Science References. doi:10.4018/978-1-60566-190-2.ch009

McNamara, T. (2008). *Language testing*. Oxford, UK: OUP.

Meskill, C., & Quah, J. (2013). Researching language learning in the age of social media. In M. Thomas, H. Reinders, & M. Warschauer (Eds.), *Contemporary computer-assisted language learning* (pp. 39–54). London: Bloomsbury Academic.

Mesthrie, R., & Bhatt, R. (2008). *World Englishes: The study of new linguistic varieties.* Cambridge, UK: CUP. doi:10.1017/CBO9780511791321

Olek-Taszarek, W. (2014). Dinosaur time travel: The way and how of ICT in a language classroom. In M. Krawiec (Ed.), *Cross-curricular dimensions of language learning and teaching* (pp. 47–60). Cambridge Scholar Publishing.

Pereira, D., Fillol, J., & Moura, P. (2019). Young people learning from digital media outside of school: The informal meets the formal. *Comunicar, 58*(27), 41–50. doi:10.3916/C58-2019-04

Rogers, A. (2008). Informal learning and literacy. In M. Stephen & H. Nancy (Eds.), *Encyclopedia of language and education* (2nd ed.; pp. 133–144). New York: Springer. doi:10.1007/978-0-387-30424-3_41

Rogers, A. (2014). The classroom and the everyday: The importance of informal learning for formal learning. *Investigar Em Educação, 2*(1), 7–34.

Rüschoff, B. (2009). Output-oriented language learning with digital media. In M. Thomas (Ed.), *Handbook of research on Web 2.0 and second language learning* (pp. 42–59). Hershey, PA: Information Science References. doi:10.4018/978-1-60566-190-2.ch003

Sefton-Green, J. (2004). *Literature review in informal learning with technology outside school. A NESTA Futurelab series report 7.* Bristol, UK: Futurelab.

Sharples, M., Arnedillo-Sanchez, I., Milrad, M., & Vavoula, G. (2009). Mobile learning: Small devices, big issues. In N. Balacheff, S. Ludvigsen, T. Jong, A. Lazonder, & S. Barnes (Eds.), *Technology enhanced learning: Principles and products* (pp. 233–249). Heidelberg, Germany: Springer. doi:10.1007/978-1-4020-9827-7_14

Sockett, G. (2011). From the cultural hegemony of English to online informal learning: Cluster frequency as an indicator of relevance in authentic documents. *ASP, 60*(60), 5–20. doi:10.4000/asp.2469

Sockett, G. (2014). *The online informal learning of English.* London: Palgrave Macmillan. doi:10.1057/9781137414885

Swain, M. (1985). Communicative competence: Some roles of comprehensible input and comprehensible output in its development. In S. M. Gass & C. G. Madden (Eds.), *Input in second language acquisition* (pp. 235–253). Rowley, MA: Newbury House.

Travis & Joseph. (2009). Improving learners' speaking skills with podcasts. In M. Thomas (Ed.), *Handbook of research on Web 2.0 and second language learning* (pp. 313-330). Hershey, PA: Information Science References.

Trinder, R. (2017). Informal and deliberate learning with new technology. *ELT Journal, 71*(4), 401–412. doi:10.1093/elt/ccw117

Wang, S., & Heffernan, N. (2009). Mobile 2.0 and mobile language learning. In M. Thomas (Ed.), *Handbook of research on Web 2.0 and second language learning* (pp. 472–490). Hershey, PA: Information Science References. doi:10.4018/978-1-60566-190-2.ch025

Wang, Y. (2014). Acquiring the English Language through Web-based role-playing simulations: Informal Learning in Interactive Environments. *International Journal of English Language Education, 2*(1), 100–110. doi:10.5296/ijele.v2i2.5932

Chapter 2
A Guide for Mobile-Assisted Language Learning in Informal Settings:
Pedagogical and Design Perspectives

Emine Şendurur
Ondokuz Mayıs University, Turkey

ABSTRACT

The evolution of mobile devices, web technologies, and learning paradigms has been shaping MALL practice. Today, informal learning has become more important than ever since the number of resources increased and the interaction styles became multidimensional. Computer-assisted language learning was converted into different forms, and mobile-assisted language learning is one of them. In its early history, the pedagogical dimension of MALL was criticized for being very behavioristic. However, the collaborative and creative nature of Web 2.0 tools contributed to methodological changes in the MALL tradition. Although there are many mobile applications designed to improve language proficiency, cloud-based tools and other mobile applications can feed the language skills informally if the designers benefit from big data.

INTRODUCTION

The notions of teaching and learning have been shaping in line with the improvements of available technologies. Since the days when the classrooms were dominated by blackboards, textbooks, and notebooks, the meaning of instructional media has

DOI: 10.4018/978-1-7998-2116-8.ch002

Copyright © 2020, IGI Global. Copying or distributing in print or electronic forms without written permission of IGI Global is prohibited.

changed dramatically. In addition, the way people access information, interact with resources, and utilize instructional tools have been transformed into different forms. For example, before the Internet, encyclopedias and books were highly demanded as primary sources. Today, the majority of people owing to the convenience issues prefer free online sources. At that point, many different issues including ethics, reliability, currency, and accuracy of the source emerge. Nevertheless, the use of modern tools is inevitable for modern people (Lebo, 2016). That's why the issues about technology integration are still quite popular among scholars.

The concept of instructional media sometimes can be fuzzy although the modern media have distinctive features. Although the issue of whether teachers are instructional media or not is arguable, recently the discussion has a new direction towards how people or any other source can serve as instructional media regardless of environmental settings (Looi et al., 2012). In other words, the new educational paradigm considers the notion of learning free from boundaries. The following aspects exemplify the industrial paradigm: sorting students, teacher-centered instruction, presentation of information by teachers, time-based progress, standardized instruction, norm-referenced testing, individual working skills, and unpleasant activity (Reigeluth, 2016). On the other hand, those aspects have turned into other forms by post-industrial paradigm. The following features can be listed as the main issues of the current paradigm: learning-focused, learner-centered instruction, learning by doing, attainment-based progress, customized instruction, criterion-referenced testing, collaborative skills, and enjoyable activity with life-long learning (Reigeluth, 2016). Unlike the demands of industrial age, the needs of information age are multidimensional, and thus the frames of reference for learning have been changing. In classical taxonomy of Bloom, the knowledge was an important sign of cognition, but owing to the paradigm changes, the taxonomy was revised (Krathwohl, 2002). Knowledge was converted into remembering because rote memorization does not meet the demands of current paradigm. It is more appreciated if one remembers where to look, how to find, and how to do something. People ask Google and then access any kind of information; therefore, creating something innovative and unique is the ultimate goal of today's education.

The post-industrial paradigm brought about redefined roles of both teachers and learners, but technology was also equipped with new roles for learning (Reigeluth, 2016). The older technology was lack of customization, but modern ones keep approaching to science fiction movies thanks to artificial intelligence. Record keeping, planning, instruction, and assessment for student learning can be considered as new roles of technology in the current paradigm (Reigeluth, 2016). Another dimension of education that has evolved is related to the transactional distance between the learner and the teacher. The trends in instructional technologies also affected the way learners and teachers interact (Reiser, 2012). Once upon a time, distance education was taking place with the help of letters, but today it welcomes unimaginable forms of interactions. At present, people learn anything regardless of time and place thanks to the Internet-based technologies. The increased number of distance education centers, rapidly evolving free courses offered by either formal or non-formal institutions, highly rated Massive Open Online Courses (MOOC), or millions of mobile applications on the market can be named as signs of evolving learning environments. In other words, the learning experiences of students are not limited to classrooms, teachers, books, course hours, or any other aspect of regular learning settings (Barlow, 2008; Looi et al., 2012).

This chapter aims to increase awareness of potential situations to benefit from mobile language learning tools in informal settings. There are lots of MALL applications in the market. They appeal to a wide range of learners. They can be used for formal or informal learning purposes, but as the mobile application markets do not edit the content or design of apps, their pedagogical aspects might become questionable. The current century requires a group of digital skills to be possessed by any learner, and thus digital literacy can emerge in any context. The informal language learning happens in a non-planned manner, which means the learner is in charge of managing the flow. The guidance (or control) of the instructor might prevent the misusage of the apps in formal settings, but informal settings are more vulnerable to unexpected outcomes. That is why closing the gap between the formal and informal usage of MALL tools is a valuable issue to be discussed. The chapter discusses the potential integration of the available tools through exploring the latest challenges of technology. Moreover, the chapter presents a wider perspective containing pedagogical, technological, and design aspects of current and upcoming tools. Although the main focus is on MALL, other opportunities including Web 2.0, Web 3.0, and Web 4.0 are also mentioned in detail. Moreover, certain examples are presented in order to make clearer the future integrations of Web 3.0 and Web 4.0. In addition to pedagogical integration, this chapter also discusses the user experience practices of MALL tools. As the accessibility of mobile devices increased in education, the number of MALL tools increased. In such online markets as AppStore, one can search and find various types of apps to be used. There are very successful apps such as DuoLingo, but some of them have lots of problems in terms of usability. These problems are discussed in terms of pedagogical, technical, and design perspectives of user experience research. Good examples of current practices are also discussed. At the end of the chapter, a list of practitioner guides is provided.

MALL in the Information Age

In formal learning experiences, the learner is generally bound by time and place to learn. The courses are planned according to a curriculum, and there is no room for flexible options such as selecting the instructional materials to be studied. However, informal learning scenarios are more convenient for flexible applications, especially the ones with mobile opportunities. Although teaching and learning processes had been perceived as restricted with school buildings for a long time, the information age paradigm had raised the value of out-of-school experiences as well as in-school but out-of-curriculum experiences (Reigeluth, 2016). Based on this assumption, the three types of learning can be listed as follows: formal, non-formal, and informal learning. The formal learning can be associated with mainstream education. It occurs in regular school systems, and hardly goes beyond school walls. These walls can be virtual in electronic learning (e-learning), mobile learning (m-learning), or digital learning (d-learning) context, but still exist regardless of overall context. Such type of learning has certain curricula to be followed with certain outcomes resulting in earned degrees or diplomas. The non-formal learning is frequently confused with the informal learning, but it "takes place alongside the mainstream systems of education and training and does not typically lead to formalised certificates" (European Commission, 2000, p. 8). Unlike formal one, non-formal learning can be both more individualized and flexible. Attending a course to improve language skills, participating in workshops about painting, following courses offered by online content developers such as Udacity, CodeAcademy, etc., and many other online or face-to-face examples represent non-formal learning phenomena. The informal learning, unlike others, "is a natural accompaniment to everyday life" (European Commission, 2000, p. 8), and can be experienced in many different forms, which can be even non-conscious. The other types have their motivation from intentions such as having a degree, a certificate, or improving skillsets in addition to be conscious about the outcomes, whereas informal learning can be completely non-conscious and unintentional, which makes it unique.

The digital technology has enormous potentials to afford any type of learning, especially the informal one, because there are lots of virtual stimuli in addition to real world. Besides conscious efforts to regulate digital teaching-learning environments, there is limitless number of uncontrolled visual or auditory elements all around the digital space. No matter how a learning management system is designed in accordance with universal design principles, appropriate pedagogical and content approach, and technical standards, it may not guarantee a distance learner's engagement in course-free content, such as checking inbox for new e-mails. In this case, the learner may not be aware of learning from that e-mail content, or any other piece of information from any other source (TV, phone, roommate, and so on). That is the reason why informal learning should be considered for any type of learning situation. According to Schugurensky (2000), different combinations of intentions and consciousness are the fundamentals of informal learning experiences.

The taxonomy based on this assumption offers the following four types of informal learning (Bennett, 2012; Schugurensky, 2000): *self-directed, incidental, tacit (or socialization),* and *integrative*. Self-directed informal learning is one of the highly studied topics in terms of language learning literature. In *self-directed* informal learning (Schugurensky, 2000), the learner has a certain intention to learn something and is aware of the occurrence of learning. Considering a mobile language learning situation, looking up the meaning of a word via a mobile dictionary application can be a good example for self-directed informal learning. The learner aims to find the word, and understands its meaning as s/he reads. *Incidental* informal learning (Schugurensky, 2000) is different from the former one in terms of intentions, i.e. the learning is triggered with unintentional attempts of which results in learning and the learner is aware of the occurrence of it. Assuming a person wandering around Instagram hash tags (#) and running into funny caps (a social media jargon used for visuals with added red rectangles having words, phrases, or sentences on it), then s/he learns a new word from one of those caps. It does not matter when s/he becomes aware of knowing that word. It is a common form of informal learning because people frequently ask themselves how should I learnt that song despite not tending to listen to. Social networking sites are full of opportunities to learn incidentally thanks to analytics they applied. In *Tacit (socialization)* informal learning (Schugurensky, 2000), the learner neither has intention to learn nor is aware of learning. Such kind of learning can emerge when people internalize certain rules of community. For example, considering an online language learning scenario, one can learn certain daily conversations from subtitles of watching TV series. S/he can use these sentence patterns in another context without being conscious about where, when, and how s/he had learnt. Moreover, watching the series and learning something from them were not intentional at all. In online communities, tacit informal learning can be observed frequently. Belonging to a language learner's online page on Facebook for instance may push members to follow a group of ethical rules. They may not be written ones, but members can observe, learn, and internalize, as there is a multidimensional network of interactions. The last type is *integrative* informal learning (Bennett, 2012). Although there is an intention to learn, the learner is totally unaware of the occurrence of learning. For instance, a beginner language learner may not figure out how to contact with a pharmacist to buy a painkiller and search the web to find the specific name of the pills. Conducting a web search is an intentional attempt to learn the information, but s/he may not find the exact name but learn that they have a generic name called painkillers. Then s/he can show choose the exact one among other painkillers offered by the pharmacist. In this case, the learner is not aware that s/he has learnt the word painkiller in another language.

Although designing and conducting studies of informal learning contexts are highly sophisticated, there are many recent projects reporting interesting results.

There are studies focusing on the role of technology in both formal and informal learning (e.g. Morgana, 2018), and there is a tendency to observe the effects of mobile technologies (e.g. Çakmak & Erçetin, 2018) on language learning processes. The flexible nature of mobile technologies results in frequent usage during daily lives of laymen, and therefore the concept of m-learning emerged as a new way of learning, which in turn triggers new forms of learning (Khaddage, Müller, & Flintoff, 2016). Despite lack of a universal definition, it is assumed that m-learning is different than other forms of learning with regards to flexibility (Peters, 2007), communicational aspects (Sharples, Taylor & Vavoula, 2010), sociocultural features (Kearney, Schuck, Burden, & Aubusson, 2012), and situated learning opportunities (Gikas & Grant, 2013).

From the days of personal computers (PC) in the 1980s to personal smartphones in the 2000s, there has been a dramatic change in the technology integration process. Once the high reputation schools were equipped with PCs, today the schools associate their reputation with STEM laboratories surrounded with robots, tablet computers, and many other high-tech gadgets. That is why the similar change in language learning is inevitable. As a well-known researcher, Skinner's programmed instruction movement and the famous teaching machines can be considered as milestones for educational technologies (Dricoll, 2005). The specifically designed machines allow the learner to lead the way thanks to following features: step-by-step flow, pacing in student's own speed, providing overt responses, and immediate feedback (Reiser, 2012). This movement prevailed for a long period, and shaped its successors, and therefore some researchers associate this period with behaviorist paradigm (e.g. Bax, 2003). In early history of computers, small steps with immediate feedbacks were accompanied with drill and practice methodology, which is very common in either Math or language learning. Based on this aspect of Computer Assisted Language Learning (CALL), Warschauer (2000) named it the structural period. These periods of computer technology were quite restricted in terms of interaction. Most of the users were highly educated, but since the commencement of PCs, the way people interact with computers dramatically changed during the 1980s. Graphical user interfaces expanded the number of target users and computers became more affordable by the laymen. These periods were important for CALL because there are lots of new versions of it. English tutoring CDs/sets, relatively rich formats of media, e-encyclopedia and dictionaries, relatively smaller file sizes, and increased storage capacities can be named as some of the characteristics of that period.

The computer as an instructional tool is already very much different than the previous instructional media traditions (Schrum & Levin, 2009). Beyond its affordances to learning, it changed the way people communicate with the integration of Internet technology. Synchronous communication became easier. The barriers of learning at a distance did not exist any longer. Physical distance and time have gained new meanings for people. The increased quality of multimedia and hypermedia opened the way for new learning methodologies. The PCs have transformed into more portable forms, and finally the phones are converted into smart ones. The Internet has also evolved into more convenient forms. From the Web 1.0 through the Web 4.0, many aspects have changed. For example, Web 1.0 was generally known as static web referring to its read only feature, however, Web 2.0 enabled any user to become author of any web service, and thus it is no more static. Although it was a good opportunity for laymen to exchange information, accessing the reliable and valid information became harder due to the lack of control by authorities. Someone can write about any topic, even s/he has no idea at all, and then the content can travel around the world. Indeed, that is the way fake news or Internet frauds spread quickly. On the other hand, Web 2.0 tools offer tremendous amount of experiences for the world of education. These will be explained and discussed in detail under MALL Tools title. The introduction of Web 3.0 increased the features of web with high semantic aspects generally associated with artificial intelligence. In other words, machines and things can communicate with each other and learn either to create data or to benefit from each other's database.

As one searches the web, the visited sites learn what the user looks for, and then provide with advertisements or suggestions on other places user visits. It is not a weird situation if a learner using any mobile learning app comes across an ad offering discounts on previously interested item on a shopping site. Although the contexts are different, the user is the same, so his/her interests, too. There are limitless opportunities of Web 3.0 to be integrated into informal learning scenarios. For instance, learning analytics can be designed in a manner to support informal learning scenarios. Web 4.0 is relatively a newly emerging era and associated with virtual realities. The web is getting smarter, so the scenarios in science fiction might become real very soon. Considering a language learning case within Web 4.0 era, the smartphone might detect that the user has pronunciation problems with a few words as a result of listening to his/her business conversations with foreigners. Then it decides to play songs/videos having the most frequent usage of those words, and then as the pronunciation problem diminishes, the support vanishes. People actually experience similar scenario owing to big data analytics as a part of Web 3.0, and the contributions of Web 4.0 structure can move the scenario beyond imaginations. In short, it is known that the technology is not the panacea of all language learning problems, but the pedagogical approaches and practices might be affected and shaped by technological advances (Al-Kadi, 2018). Table 1 summarizes the important points about Web and language learning.

Table 1. Evolution of Web and Language Learning

	Web 1.0	Web 2.0	Web 3.0	Web 4.0
Role of the learner	-Reader -Commentator -Consumer	-Content Generator -Consumer + Producer	-Consumer + Producer	-Consumer + Producer
Role of the technology	-Static -Providing with read-only content -Mostly asynchronous communication	-Dynamic -Providing with writing opportunities -Modification of content -Highly interactive	-Providing with suggestions based on track of users kept by databases -Semantic -Machine-to-machine interaction -Real-time information -Clouds	-Offering solutions -Intelligent -Compatible across platforms (web desktop)
Content	Generated by web publisher	Generated by web publishers + users	Generated by web publishers + users + machines	Generated by web publishers + users + machines + intelligent agents
Language Learning Tools	-Dictionary websites -Quiz and exercise web site -Forums	-Social networking sites -Blogs -Wikis -Quizlet -Kahoot	-Applications utilizing semantic database (e.g. search engines) -Duolingo -Rosetta Stone -Fluenz -Google docs -Google translate	-Smart personal assistant -Context-aware apps -L2-TOR Project (Belpaeme et al., 2015) -WebOS (e.g. EyeOS) -Google assistant
Pedagogical Approaches	-Drill and practice -Rote memorization -One-way feedback -Mostly behaviorist -Teacher-led	-Project-based -Problem-based -Gamification -Collaborative -Creative -Mostly constructivist -Learner autonomy	-Virtual teaching -Collaborative -Scaffolding tools -Ubiquitous learning -Eclectic -Learner autonomy -Semantic guidance	-Cyber instructors -Collaborative -Distributed learning -Eclectic -Learner autonomy -Intelligent guidance -Authentic

Using MALL in Informal Settings

CALL is now converted into MALL with variations in practice such as Smartphone Assisted Language Learning (SPALL). In its early periods, studying with mobile devices had many limitations including screen size, screen resolution, power capacity, multimedia capacity, and so forth. Nevertheless, most of them have been increased in quality, which resulted in better user experience. The current studies show positive effects of MALL on learning (Flores, 2015; Ketyi, 2016; Leis, Tohei, & Cookie, 2015; Rosell-Aguilar, 2018; Shadiev, Hwang, & Liu, 2018). In a recent review, Hwang and Fu (2019) reported some significant changes in the field of mobile language learning research. Some of them can be listed as follows:

- The recent studies have a tendency towards issues related to multiple language skills rather than individual skills alone;
- There is a methodological transformation into mixed paradigm;
- The number of longitudinal studies has been increasing;
- The positive effects on such skills as writing and speaking were reported frequently;
- The number of studies stating the effects on such skills as listening and reading is limited (Hwang & Fu, 2019).

The inclusion of technologies for interventions might be in danger of novelty effect, which increases the need for longitudinal studies. The positive attitudes or significant results should be interpreted cautiously. The literature includes considerable amount of studies reporting positive effects, but there are also the ones with no significant differences. In Rachels and Rockinson-Szapkiw's (2017) study, the Duolingo was used as an intervention in comparison to regular in-class activities, and it did not resulted in differences in achievement. This study has importance due to its relatively longitudinal nature and its statistical control over beginning scores. In Dizon's (2016) study, Facebook was compared to traditional writing method. Despite the increased fluency in writing of Facebook group, no significant differences were found regarding lexical or grammatical dimensions. Benefitting from the mobile apps may require time and familiarity with the system (Osman & Chung, 2011; Zhang, 2016), and they may not be appropriate for all types of learning tasks (Brown, Castellano, Hughes, & Worth, 2012; Epp, 2017; Naderi & Akrami, 2018). Certain aspects of learners might also affect the performance while learning from the app: such as field-dependency (Chen, Chen, & Yang, 2019), gender (Liu, Zheng, & Chen, 2019), age (Lee, 2019a; Stevenson & Liu, 2019; Terantino, 2016) and level of proficiency (Barcena & Read, 2016; Liu et al., 2019). Although the majority of the studies favor the collaborative nature of MALL, Zhang (2016) pointed that discussion on a mobile app was not perceived as convenient. On the other hand, for such audience as young learners, MALL tools may require guidance by real people (Terantino, 2016).

The literature about MALL can be divided into two parts: (a) the studies integrating MALL into formal settings, and (b) the ones locating MALL into informal settings. On the other hand, recent studies tend to include informal learning as an extension of formal learning. For example, WhatsApp and search engines were used to complete assigned reading materials outside the school and the social interaction continued online (Hazaea & Alzubi, 2018). WeChat was used to increase language proficiency, and it served as an extension of formal language learning (Shi, Luo, & He, 2017). This trend might not emerge due to planning, but mobile technologies keep changing rapidly. Web 2.0, cloud-based tools, machine learning, and many other improvements bring about convenience to create multidimensional/multi-way connections on the web. No matter how strict the research design of the study is, the participants are vulnerable to any online stimulus. That is why it is hard to claim that MALL is isolated from its informal learning aspects.

Smartphones are perceived as distractors of learning in classrooms, and sometimes become competitors of instructors. In order to take advantage of such intense usage of phones, Cottam and Tat (2018) encouraged students to use mobile applications ranging from calculator to social networking sites, and they reported increased success and high engagement. Although the mentioned study consists of various observations and data sources throughout the action research design, the most significant finding was the high engagements of homework and collaborative learning processes (Cottam & Tat, 2018). The increased performance or success with regards to language proficiency including reading, writing, listening, and speaking was also reported in other studies (Azli, Shah, & Mohamad, 2018; Chen & Li, 2010; Keezhatta & Omar, 2019; Leis et al., 2015; Naderi, 2018; Shih, 2017; Osman & Chung, 2011; Sun et al., 2017). The integration of phones into language lessons can improve learner autonomy (Chen & Kessler, 2013; Hazaea & Alzubi, 2018; Leis et al., 2015; Zhang, 2016). Self-regulation and learner autonomy have become key skills to survive in the 21st century, and thus researchers focus on them in mobile language learning situations. In their study, Chen et al. (2019) integrated the self-regulation principles into a mobile language learning application and found significant effects on both learning performance and motivation.

MALL can serve as complementary learning for students having difficulties in traditional settings. Mobile apps might become substitutes of content, teacher, peer, or any other source of information. They can help to understand either unclear or missed parts of the lecture due to absenteeism, and thus they can serve as convenient tools for language learning (Azli et al., 2018; Hazaea & Alzubi, 2018; Rosell-Aguilar, 2017; Vurdien, 2017; Zaki & Yunus, 2015; Zou & Li, 2015). Even a simple Google application can contribute to development of language skills (Vurdien, 2017). In the literature, there are studies, which completely integrated mobile apps into instruction (Keezhatta & Omar, 2019; Masood & Halimi, 2019; Naderi & Akrami, 2018). In other words, mobile tools become the main media of instruction. For instance, Telegram was used to improve reading comprehension, and the instruction took place in Telegram groups (Naderi & Akrami, 2018).

It can be too shallow to consider only the benefits of learners. Instead, perspectives of the instructional designers and teachers are also important because they are the ones who accept and adopt the MALL tools. The beliefs, perceived ease of use, and the way tools serve as scaffolds can be named as key factors affecting the integration of MALL in educational settings by practitioners (Morgana, 2018; Barcena & Read, 2016). Despite being started in formal settings, these studies had all informal dimensions as they bring about collaboration out of school, which is an important dimension of informal learning (Bennett, 2012; Schugurensky, 2000). For example, in their study, Shadiev et al. (2018) compared traditional methods with a mobile multimedia learning system, and found that the collaborative nature brought about better performance besides metacognitive skill improvements. A similar comparison was performed by Shih (2017), and resulted in MALL group's higher performance.

A post on the flow of a social networking site, a cartoon listed on search engine results page, an event announcement on a website, and many similar experiences can be named as informal learning materials. They are not written in purpose of teaching a language, but still can serve as incidental learning tools. Unlike those ones, there are materials specifically designed for triggering informal learning situations. For example, Wrigglesworth (2019) used a platform to increase the interaction among students as they send multimedia messages in informal settings via smartphones. Such an approach has potential to support continuous improvement beyond school settings because mobile technologies play key roles in collaborative language learning (Brown, Castellano, Hughes, & Worth, 2012; Kukulska-Hulme & Viberg, 2018). Some creative informal materials to be used via smartphones can serve as scaffolds of language learning in different contexts. For example, Liu, Hwang, Kuo, and Li (2014) used QR codes in a fitness center to provide people with reading materials, and therefore created opportunity to build fitness terminology in English.

The learner in informal learning situations generally finds himself/herself in the middle of learning process with/without intentions or with/without awareness. It is different from formal learning, which is usually launched by instructors. Epp (2017) explored the learner-initiated MALL usage among migrants. The study is crucial to understand the real situations people apply for the guidance of mobile apps in a form of informal learning. The findings suggested that the utilization of tools remained shallow, including reviewing words, listening, translating (Epp, 2017). Considering different types of users, Terantino (2016) made use of MALL for preschool children and examined the informal learning process in collaboration with parents for six months. Unlike other studies, the participants were free to choose and use the mobile app (e.g. Busuu Kids, Spanish Smash, LinguPiguin, etc.). The findings showed that gamification features attracted attention, vocabulary and listening skills were improved, but the improvements in other language skills were not achieved by apps (Terantino, 2016).

There is a growing body of literature considering the impact of informal digital learning of English. The studies usually conclude that the learners become more confident with such language skills as speaking if they experience wide range of informal digital learning (Lee, 2019a), but the quality of it matters due to being positively related to vocabulary outcomes (Lee, 2019b).

MALL Tools

In addition to the search engines, e-books, dictionaries, etc., mobile language learning/ practicing applications are quite popular in any MALL situation. It can be challenging to differentiate their purpose of usage, i.e. it may not be very clear if they serve as non-formal or informal tools. At that point, the intention and consciousness should be analyzed to make the type of learning explicit. In addition, they can be even a part of formal learning scenario as well (Steel & Levy, 2013). Duolingo, Busuu, Babbel, Italki, and Mondly are some of these popular apps. They have millions of downloads and their users show a wide range of demographical varieties. In the literature, it can be inferred that learners favored such applications. The possible reasons can be summarized as follows:

- Gamification (Deterding, Sicart, Nacke, O'Hara, & Dixon, 2011; Flores, 2015; Masood & Halimi, 2019; Munday, 2016; Şendurur et al., 2017; Terantino, 2016),
- Simple user experience and convenience (Azli et al., 2018; Flore, 2015; Munday, 2016; Shih, 2017; Şendurur et al., 2017; Vurdien, 2017),
- Reliable tool (Rosell-Aguilar, 2018),
- Improved vocabulary (Chen & Li, 2010; Masood & Halimi, 2019; Rosell-Aguilar, 2018),
- Improved reading skills (Keezhatta & Omar, 2019; Naderi, 2018),
- Social influence (Botero, Questier, Cincinnato, He, & Zhu, 2018),
- Enhanced learning process or language proficiency (Azli et al., 2018; Shi et al., 2017),
- Increased learner autonomy (Chen & Kessler, 2013; Hazaea & Alzubi, 2018; Zhang, 2016),

- Serving as a complementary tool for language learning (Finardi, Leao, & Amorim, 2016; Morgana, 2018; Şendurur et al., 2017),
- Elimination of communication barriers (Epp, 2017),
- Improved oral skills (Sun et al., 2017).

Mobile Apps with Web 2.0 and Cloud Technologies

There are many varieties of mobile applications under language learning/practicing themes. One of the prevalent application themes is *dictionary*. Their interfaces, designs, and services show differences, but learners' motivations to use them can be categorized into three: customization, learning, utility (Liu et al., 2019). A dictionary application such as Merriam-Webster, which is free, offers many customization options such as saving the history of recent searches or favorite words, and so on. For any user, customization gives users freedom of choice, which is very important for satisfaction of users. The flexible dictionary apps can serve as learning tools. For example, when the learner hears an unknown word without knowing how to spell it, the dictionary may provide a list of possible words, which may be a great opportunity for informal learning. In terms of utility feature, a dictionary app can be viewed completely as a tool to leverage the success. For example, assuming a beginner level learner having limited vocabulary skills, a dictionary app can help him/her to distinguish in what context the word can be integrated into a sentence in a meaningful way.

Portable translators have been available at the market for a long time. Although their formations have changed, the expectations of learners remained the same. In other words, they expect translators to serve as simultaneous interpreters. The process of translation is really complicated. The real meaning of the word *bug* in a sentence depends on its context. It may be either an error in a computer program or a real *bug* coming from the animal world. Considering the importance of context, Myers (2000) reported a group of benefits of using personal digital assistants (PDA) in translation, but also pointed out the necessity of similarity between translated languages. This challenge is aimed to be resolved as Web evolves through the Web 4.0, but now the problem still exists. *Google translate* runs properly and keeps improving with the inclusion of augmented reality and natural language processing. There are studies showing the potential academic benefits of Google translate despite its deficiencies (Balk, Chung, Chen, Trikalinos, & Chang, 2013; Groves & Mundt, 2015). It fits well with informal learning. For example, a tourist may shop at a local market by scanning the unknown words in a foreign language via smartphone and translate it immediately, i.e. there is no need for typing or spelling the words, and the camera will identify the word.

Listening materials constitute an important part of language learning experience even though some of them are not designed for instructional purposes. *Podcasts* can both enrich the listening opportunities outside the school and enhance the learning opportunities. *Instant messaging* (e.g. ICQ and MSN Messenger) was a fantastic way to communicate in the early stages of the Web. Today, such new apps as WhatsApp were built upon basic features of instant messaging, but their practical contributions went beyond expected usage, and hence they turned into social networks. WeChat is one of the examples of instant messaging apps with a range of features including voice messaging. It was integrated into a language learning scenario by Shi et al. (2017), and the results indicated many gains in favor of language proficiency.

The number of *mobile language learning apps* has been increasing in the market. Duolingo, Rosetta Stone, Fluenz, Quizlet, Babbel, Busuu, and Yabla are some of the popular ones. The studies integrating these tools into research design investigate various dimensions ranging from learning to motivation. For example, Botero, Questier, and Zhu (2019) utilized Duolingo to have insights of students' informal learning experiences. They reported that the usage levels reached the peak during holidays, but the way learners use the app do not match with their thoughts about the app (Botero et al., 2019). The needs of learners may vary throughout the academic year. The range of benefits may decrease as they have busy exam schedules, whereas practicing through a tool like Duolingo may turn into a free time activity. Duolingo is a free app, but there are many others having a variety of features. In addition to these commercial apps, there is also non-benefit apps developed by researchers (e.g. Barcena & Read, 2016; Gaved & Peasgood, 2017; Keezhatta & Omar, 2019). Edmodo, ARIS, MondlyAR, ActionBound, Scratch, Unity, etc. are some other tools that have not general purposes, but can be used by language learners or instructors.

When Web 2.0 entered into ordinary people's lives with compatible mobile features, the way learners manage the content had changed. They were not only the readers of the content but also the writers of it. Thanks to *social networking sites* it became possible to share any content quickly and easily. Facebook, Instagram, Twitter, and Youtube are some popular sites. Although they were not originally designed for learning purposes, educators keep benefiting from them in various ways (e.g. Dizon, 2016). In addition, these tools provide great opportunities to support informal learning. For example, Hazaea and Alzubi (2018) conducted a study to observe how students utilized WhatsApp and search engines as an extension of formal reading class. The authors claimed that the included apps helped to improve learner autonomy because learners were in charge of decisions about materials, time, location, etc. The integration of social networking sites can even contribute to kids' language practice. There are a few mobile social networking apps appealing to kids. Papa, Sock Puppets, Kids for Narration, and Speaking English Fluently are some of them, but according to Sun et al. (2017) the first one is the safest for kids. That is why they conducted their study via Papa app. They observed that first graders' oral fluency was improved (Sun et al., 2017).

There is a group of Web 2.0 tools frequently preferred by language users. Since the Web 2.0 enables users to create and share their own content, the informal learning opportunities are limitless when accompanied by mobility feature of devices. The number of Web 2.0 tools is increasing day by day, but a three-dimensional taxonomy can clearly summarize the tools and its features (see Table 2). The table also offers potential integration of tools to informal language learning situations.

MALL from User Experience (UX) Perspective

Pedagogical Aspects

The pedagogy behind MALL is not very much sophisticated, but it confronts with the curse of instructional media. That is, any new medium was expected to boost learning magically by substituting the previous ones, but the history of instructional media includes many disappointments (Mayer, 2009). On the other hand, this miracle has never happened so far. The practitioners and researchers had learnt that the novelty effect emerges when the medium is new to learners. Moreover, the history of educational technology showed that any new technological tool is just a tool, so it may not make difference unless appropriate pedagogical approach is integrated into the process. The studies or practical approaches are vulnerable to fall into this pedagogical challenge through a greater focus of the technology itself. In fact, MALL has great potential to adopt collaboration, authentic tasks, problem-based situations, and many other constructivist approaches.

Gamification is a popular approach utilized in a variety of fields. It sets clear goals and tasks to be completed in a challenging way. Educators appreciate keeping motivation of learners at high levels to ensure learning. In formal settings some factors affecting motivation can be manipulated to an extent, but in such informal settings as mobile apps, the number of distractors may not be predicted due to online activity at other online platforms. In other words, Web 3.0 can either result in distractions or contributions. On the other hand, gamification adds value to MALL's pedagogical dimension. In Masood and Halimi's (2019) study, the gamification features of Memrise application were observed, and the results indicated that learners were motivated to use the app owing to the gamification elements including selectable modes, level variety, points to be collected, flower growth, and progress report. The elements of gamification certainly strike attention of younger students, too.

Assessment and evaluation is one of the challenging structures of pedagogical approaches. When the context is mobile and informal, it may become valuable to collect data in order to understand learners' progress in language learning. Although,

Table 2. Taxonomy of Web 2.0 Applications (adapted from Orehovacki, Bubas, & Kovacic, 2012) & Their Potential Integration of Informal Language Learning

<table>
<tr><th>Dimension</th><th>Sub-dimensions</th><th>Tools</th><th>Informal Language Learning</th></tr>
<tr><td>Type</td><td>Wiki, microblog, blog, social network, social bookmark, mashup, podcast, e-portfolio, virtual world, note-taking & editing services, office applications, source code publishing services, mind mapping & flowcharting services, creative learning applications, multimedia publishing services, audio & video forums, video conferencing services.</td><td rowspan="3">-Wikipedia
-Blogger
-Twitter
-Facebook
-CiteULike
-Atom
-iTunes U
-ScreenCastle
-Mahara
-Second Life
-SpringNote
-Google docs
-GitHub
-Bubblr
-Youtube
-Instagram
-Slideshare
-Userplane
-Bubbl.us
-Kahoot</td><td rowspan="3">-Personal blogs as informal writing exercises.
-Following updates of famous foreigners on social networking apps.
-Personalizing the reading feed via mashups.
-Recording own video/audio and sharing them.
-Listening to podcasts released by native speakers.
-Attending a virtual community or class in virtual worlds.
-Creating cartoons to simulate funny conversations.
-Following multilingual newsagents' pages.
-Contributing to a wiki content.
-Leaving comments in travelling apps.</td></tr>
<tr><td>Function</td><td>Collaboration, sharing, communication, knowledge organization, learning support, artifacts integration.</td></tr>
<tr><td>Cognitive Processes</td><td>Remembering, understanding, applying, analyzing, evaluating, creating.</td></tr>
</table>

most of the apps are able to keep track of students' learning, the smart systems connecting informal learning apps to formal learning management systems are not very common, yet. On the other hand, mobile apps specifically designed for assessment can lead the way to increase such an appreciated connection from formal to informal MALL practice. For example, Titova and Samoylenko (2017) observed that the enquiry-based approaches and improved opportunities for peer learning are appreciated by learners. For example, a course management system can be fed by data streaming from a social networking group, and can decide on misconceptions of students, and then generate content to eliminate them.

MALL has certain pedagogical challenges. Burston (2014) raised some issues as follows:

- MALL still sticks to behaviorist approach with regards to content delivery.
- A systematic monitoring of student performance is restricted, and therefore the feedback mechanisms are very limited.
- Transmission model of instruction dominated the attempts to personalize learning environments.

There have been many improvements of web technology since Burston (2014) had listed these challenges. Today, as the mobility features improved, learner-oriented pedagogies were supported. Thanks to big data, monitoring a learner performance whether formally or informally is possible with personalized feedback. Virtual bots can initiate conversations, and offer new content based on improvements of learner. Moreover, location-based applications, context-aware features are embedded into apps (e.g. Chen & Li, 2010; Gaved & Peasgood, 2017). On the other hand, apps should renew their pedagogical approaches to eliminate additional external guidance such as parent of younger kids (Terantino, 2016).

In addition to pedagogical challenges, the utilization of MALL tools in both formal and informal learning might bring about certain negative outcomes, which should be seriously considered by practitioners. For example, digital games can serve as informal learning tools to improve conversation skills and vocabulary knowledge (Rankin, Gold, & Gooch, 2006), but as those games are highly engaging, there is a danger of addiction if the learner loses self-control (Lewis, 2016). Moreover, spending too much time with mobile devices may result in health-related problems. Learners can experience eye-fatigue, physical pain, headache, and many other problems, which are very similar symptoms of addictions (Kim & Kim, 2010). Another negative outcome might be the premise of reliable information, which can result in fragmented learning, misconceptions, or unlearning. Split attention and/or redundancy effects may hinder the smoothness of learning process. The user may build wrong connections, and the constructs learned in formal learning might be affected. Finally, integrating MALL into educational (whether formal or informal) settings might need special attention since the utilization of such tools make users vulnerable to attention deficits or hyperactivity disorders (Seo, Kim, & David, 2015).

Technical Aspects

The history of mobile devices dates back to the 1990s. Personal Data Assistant (PDA) and laptops introduced a new dimension to available computer technologies. Today, the mobile devices have reached to wearable versions such as smart watches. A brief timeline is introduced in Figure 1.

The lighter devices, better screen resolution, high processing capability, and many other technical amendments affected user experience in a positive way. On the other hand, the need for extra stationary is still a technical challenge for MALL context, because the lack of it can hinder the anytime-anywhere premise (Burston, 2014). Moreover, the compatibility across platforms is one of the biggest technical challenges for MALL (Burston, 2014), but the dramatic evolutions of web technologies started eliminating these challenges. For example, the learner can use Google docs regardless of the technical differences, i.e. the type of operating system (Windows, MacOS, Android, Ubuntu, etc.) or the type of device (desktop, mobile, etc.) do not matter. What the user needs is the Internet access and a web browser.

Some technical aspects make certain mobile applications unique. For example, Duolingo has high reputation among others owing to its automatic speech recognition in addition to natural language processing. This helps to increase learner satisfaction by providing flexible user entry, and hence keeps the flow of the learning at an optimum level (Gafni, Achituv, & Rahmani, 2017). Another example providing flexible usage is Google translate mobile app. It translates any text on any physical or virtual item into a preferred language. The control is at the user's hands thanks to augmented reality feature of the app.

Besides constant changes in devices, the networking technologies keep changing, too. For example, sending a message with multimedia features was not possible via cellular phones three decades ago. Today, smartphones and networking opportunities allow a wide range of processes such as communicating via Clouds. In the telecommunications market, the 3rd Generation (3G) technology can be considered as a milestone due to the considerable speed of data transfer, image transfer, and location-based tracking ("What is 5G?, n.d.). User experience got better with the affordances of 4G, but recently 5G has been preparing to expose newer experiences with better integration capabilities (Ezhilarasan, & Dinakaran, 2017). For example, data transfer will become faster than ever with 5G, and therefore integration of such

Figure 1. History of mobile devices
(Source: The content is adapted from https://www.computerhistory.org/timeline/computers/)

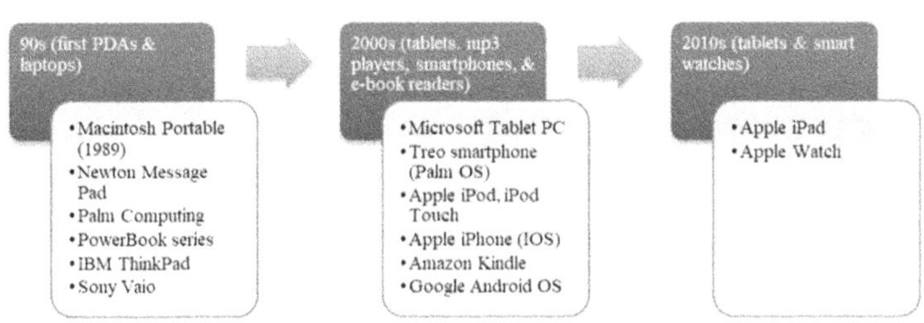

aspects as augmented reality will contribute to the educational practice ("What is 5G?", n.d.). Considering a language-learning context, the MALL tools can offer richer experiences such as smarter assistance for mispronunciations. Through the integration of virtual reality aspects, the MALL tools can also convert the real world into a virtual café where the user has to order a coffee. In other words, the authentic learning can become easier with the allowances of 5th Generation.

Design Aspects

Interfaces specifically designed for language learning are generally studied within the boundaries of cognitive theory. In a report of language learning research trends, such skills as writing and pronunciation seem still popular in studying the effectiveness (Hwang & Fu, 2019). They are generally satisfying to provide access to information, but when it comes to deeper cognitive process such as creation, the design lacks certain dimensions despite the technical capabilities. The user might use a translator app to understand a prospectus. If the translation of medical terms is not supported, it means that the tool has no support for leveraging the cognition of the learner. That is why context-aware designs should become prevalent with defined standards in the near future. There are many examples revealing visual design clues of MALL. One of them is to explore the effects of gloss type designed in three different approaches including single channel (textual only; pictorial only), dual channel (textual-plus-pictorial), and control condition (no glosses) (Çakmak & Erçetin, 2018). The results revealed that adding glosses might enhance vocabulary recognition and production, but the type of it had no effect (Çakmak & Erçetin, 2018). In addition, the users may demand traditional dictionary approaches embedded into apps (Stevenson & Liu, 2013).

Keep It Simple and Stupid (KISS) is a generic rule for designers. Google search engines home page is a representative example of this rule. It eliminates many usability issues and increases satisfaction in terms of interface design. Aesthetic features of the visual design elements can shape the emotions and attitudes towards the use of app. For example, Duolingo's design elements trigger positive feelings such as joy (Carvalho & Oliveria, 2017). The professional look of the app is crucial for users as well as reputable content (Stevenson & Liu, 2013). An important point in design is to know about the audience because the novices need more guidance in comparison to experts. For example, small bites of information can be included when the user asks for. The available skills and needs of users are quite important. The tools can offer very rich functionalities, but they become useless if the user does not benefit from them due to lack of awareness (Mason & Zhang, 2017). For example, the app can allow skipping the basic levels with short quizzes. This function should be made visible in an informative fashion; otherwise the user can leave the app due to being bored.

Practitioner's Guide

The evolutions of mobile devices, web technologies, and learning paradigms have been shaping MALL practices. Today, informal learning has become more important than ever since the number of resources increased and the interaction styles became multidimensional. As a result, these improvements bring up many issues for discussion. The following guidelines with short explanations were listed to highlight some of the issues.

- One should be cautious when planning to integrate MALL into informal settings because they are hardly isolated from formal learning practices. For example, a WhatsApp group of the class can be active during the formal course hour and one should consider to benefit from this situation.
- For beginner, the inclusion of glosses or simple translation tools may add value to usability of the application. It can be designed in any form (textual or pictorial or both). The key point is to consider cognitive load theory guidelines in order to avoid confusions about instructional messages. As the level of proficiency increases the scaffolds can be replaced into different forms with fading if necessary. Since the Web is getting smarter, the scaffolds can be shaped with the help of analytics.
- Learning management systems are preferred in many higher education institutions. Social networking and learning analytics can be adapted to those systems to support out-of-school experiences. In this way, language learning, especially the one with mobile assistance, may continue along with learners' daily routines.
- Pedagogical approaches utilizing gamification elements such as levels can provide with rich user experiences.
- Technical aspects of mobile applications as well as devices should be taken into consideration in order to prevent loss of motivation. For example, shared content should not be too large. One may consider not to spend such a long time for that content, so the loss of motivation or interest can happen. Small bites/steps/sneak peeks/text might be attractive for mobile learners in informal settings.
- The practitioners should keep in mind that content and context matter. A mobile app constantly requiring listening and speaking activities may not be convenient for learners who engage in MALL during a long commute.

- MALL tools should go beyond cognitive support. They may enrich their way to provide support for social, cultural, emotional, and metacognitive needs of learners. Machine learning can be used to enhance the available tools.
- The designers should keep the design as simple as possible by adding faded support for novice learners.
- The social networking apps bring many benefits of informal learning together. That's why, the formal learning settings should benefit from these already accepted features of them, and then remove the boundaries between school walls and virtual walls.
- The need for external guidance should be eliminated as much as possible. In other words, MALL should be smarter than ever.
- Credibility of the content might influence the learner's motivation to embrace MALL.

In conclusion, the trends and paradigms in the 21st century requires certain skill sets ranging from collaboration to creative thinking. In order to improve these skills along with the lifelong learning, one may need universal tools such as foreign languages. Thanks to rapidly improving Web and networking technologies, laymen can reach a variety of online content in any language. Mobile applications can both enrich the user experience and enhance learning. From teaching machines to m-learning, the pedagogical approaches have changed towards a more learner-oriented manner, which brought about blurring boundaries of learning settings. Practitioners and researchers should keep studying better ways to benefit from informal learning opportunities from MALL's frame of reference.

REFERENCES

Ahmed, M. D. (2015). Can smartphones pave the path towards EFL competence for Saudi college students? *Education & Linguistics Research*, *1*(2), 120–144. doi:10.5296/elr.v1i2.8467

Al-Kadi, A. (2018). A review of technology integration in ELT: From CALL to MALL. *Language Teaching and Educational Research*, *1*(1), 1–12.

Azli, W. U. A. W., Shah, P. M., & Mohamad, M. (2018). Perception on the usage of Mobile Assisted Language Learning (MALL) in English as a second language (ESL) learning among vocational college students. *Creative Education*, *9*(01), 84–98. doi:10.4236/ce.2018.91008

Balk, E. M., Chung, M., Chen, M. L., Trikalinos, T. A., & Chang, L. K. W. (2013). *Assessing the accuracy of Google Translate to allow data extraction from trials published in non-English languages*. Academic Press.

Barcena, E., & Read, T. (2016). Metacognition as scaffolding for the development of listening comprehension in a social MALL App. *RIED. Revista Iberoamericana de Educación a Distancia, 19*(1), 103–120.

Barlow, T. (2008). Web 2.0: Creating a classroom without walls. *Teaching Science. The Journal of the Australian Science Teachers Association, 54*(1).

Bax, S. (2003). CALL—Past, present and future. *System, 31*(1), 13–28. doi:10.1016/S0346-251X(02)00071-4

Belpaeme, T., Kennedy, J., Baxter, P., Vogt, P., Krahmer, E. E., Kopp, S., ... Pandey, A. K. (2015). L2TOR-second language tutoring using social robots. *Proceedings of the ICSR 2015 WONDER Workshop*.

Bennett, E. E. (2012). A four-part model of informal learning: Extending Schugurensky's conceptual model. *Adult Education Conference*. Retrieved April 23, 2019, from http://newprairiepress.org/aerc/2012/papers/3

Botero, G. G., Questier, F., Cincinnato, S., He, T., & Zhu, C. (2018). Acceptance and usage of mobile assisted language learning by higher education students. *Journal of Computing in Higher Education, 30*(3), 426–451. doi:10.100712528-018-9177-1

Botero, G. G., Questier, F., & Zhu, C. (2019). Self-directed language learning in a mobile-assisted, out-of-class context: Do students walk the talk? *Computer Assisted Language Learning, 32*(1-2), 71–97. doi:10.1080/09588221.2018.1485707

Brown, M., Castellano, J., Hughes, E., & Worth, A. (2012). Integration of iPads into a Japanese university English language curriculum. *The JALT Call Journal, 8*(3), 193-205.

Burston, J. (2014). MALL: The pedagogical challenges. *Computer Assisted Language Learning, 27*(4), 344–357. doi:10.1080/09588221.2014.914539

Çakmak, F., & Erçetin, G. (2018). Effects of gloss type on text recall and incidental vocabulary learning in mobile-assisted L2 listening. *ReCALL, 30*(1), 24–47. doi:10.1017/S0958344017000155

Carvalho, M., & Oliveira, L. (2017). Emotional Design in Web Interfaces. *Observatorio, 11*(2), 14-34.

Chen, C. M., Chen, L. C., & Yang, S. M. (2019). An English vocabulary learning application with self-regulated learning mechanism to improve learning performance and motivation. *Computer Assisted Language Learning, 32*(3), 237–260. doi:10.1080/09588221.2018.1485708

Chen, C.-M., & Li, Y.-L. (2010). Personalized context-aware ubiquitous learning system for supporting effective English vocabulary learning. *Interactive Learning Environments, 18*(4), 341–364. doi:10.1080/10494820802602329

Chen, X. B., & Kessler, G. (2013). Action research tablets for informal language learning: Student usage and attitudes. *Language Learning & Technology, 17*(1), 20–36.

Cottam, C., & Tat, C. (2018). It's your call: Addressing the rapid rise of smartphone use in EFL instruction. *The Journal of the College of Foreign Languages Himeji Dokkyo University, 31*, 67–80.

Deterding, S., Sicart, M., Nacke, L., O' Hara, K., & Dixon, D. (2011, May). Gamification. using game-design elements in non-gaming contexts. In CHI'11 extended abstracts on human factors in computing systems (pp. 2425-2428). ACM.

Dizon, G. (2016). A comparative study of Facebook vs. paper-and-pencil writing to improve L2 writing skills. *Computer Assisted Language Learning, 29*(8), 1249–1258. doi:10.1080/09588221.2016.1266369

Driscoll, M. P. (2005). *Psychology of learning for instruction* (3rd ed.). Boston, MA: Allyn and Bacon.

Epp, C. D. (2017). Migrants and mobile technology use: Gaps in the support provided by current tools. *Journal of Interactive Media in Education, 1*(2), 1–13.

European Commission. (2000). *A Memorandum on Lifelong Learning*. Brussels: Commission of the European Communities. Retrieved March 20, 2019, from http://arhiv.acs.si/dokumenti/Memorandum_on_Lifelong_Learning.pdf

Ezhilarasan, E., & Dinakaran, M. (2017). A Review on mobile technologies: 3G, 4G and 5G. In *2017 Second International Conference on Recent Trends and Challenges in Computational Models (ICRTCCM)* (pp. 369-373). IEEE. 10.1109/ICRTCCM.2017.90

Finardi, K. R., Leao, R. G., & Amorim, G. B. (2016). Mobile Assisted Language Learning: Affordances and Limitations of Duolingo. *Education and Linguistics Research, 2*(2), 48–65. doi:10.5296/elr.v2i2.9842

Flores, J. F. F. (2015). Using gamification to enhance second language learning. *Digital Education Review, 27*, 32–54.

Gafni, R., Achituv, D. B., & Rahmani, G. (2017). Learning foreign languages using mobile applications. *Journal of Information Technology Education: Research, 16*, 301–317. doi:10.28945/3855

Gaved, M., & Peasgood, A. (2017). Fitting in versus learning: A challenge for migrants learning languages using smartphones. *Journal of Interactive Media in Education, 1*(1), 1–13. doi:10.5334/jime.436

Gikas, J., & Grant, M. M. (2013). Mobile computing devices in higher education: Student perspectives on learning with cellphones, smartphones, and social media. *The Internet and Higher Education, 19*, 18–26. doi:10.1016/j.iheduc.2013.06.002

Groves, M., & Mundt, K. (2015). Friend or foe? Google Translate in language for academic purposes. *English for Specific Purposes, 37*, 112–121. doi:10.1016/j.esp.2014.09.001

Hazaea, A. N., & Alzubi, A. (2018). Impact of mobile assisted language learning on learner autonomy in EFL reading context. *Journal of Language and Education, 4*(2), 48–58. doi:10.17323/2411-7390-2018-4-2-48-58

Hwang, G. J., & Fu, Q. K. (2019). Trends in the research design and application of mobile language learning: A review of 2007-2016 publications in selected SSCI journals. *Interactive Learning Environments, 27*(4), 567–581. doi:10.1080/10494820.2018.1486861

Kearney, M., Schuck, S., Burden, K., & Aubusson, P. (2012). Viewing mobile learning from a pedagogical perspective. *Research in Learning Technology, 20*(1), n1. doi:10.3402/rlt.v20i0.14406

Keezhatta, M. S., & Omar, A. (2019). Enhancing reading skills for Saudi secondary school students through mobile assisted language learning (MALL): An experimental study. *International Journal of English Linguistics, 9*(1), 437–447. doi:10.5539/ijel.v9n1p437

Ketyi, A. N. D. R. Á. S. (2016). From mobile language learning to gamification: An overlook of research results with business management students over a five-year period. *Porta Linguarum. Revista Internacional de Didáctica de las Lenguas Extranjeras, I*, 45–60.

Khaddage, F., Müller, W., & Flintoff, K. (2016). Advancing mobile learning in formal and informal settings via mobile application technology: Where to from here, and how? *Journal of Educational Technology & Society, 19*(13), 16–26.

Kim, M. G., & Kim, J. (2010). Cross-validation of reliability, convergent and discriminant validity for the problematic online game use scale. *Computers in Human Behavior*, *26*(3), 389–398. doi:10.1016/j.chb.2009.11.010

Krathwohl, D. R. (2002). A revision of Bloom's taxonomy: An overview. *Theory into Practice*, *41*(4), 212–218. doi:10.120715430421tip4104_2

Kukulska-Hulme, A., & Viberg, O. (2018). Mobile collaborative language learning: State of the art. *British Journal of Educational Technology*, *49*(2), 207–218. doi:10.1111/bjet.12580

Lebo, H. (2016). *The 2016 digital future report.* Retrieved March 20, 2019, from http://www.digitalcenter.org/wp-content/uploads/2013/06/2016-Digital-Future-Report.pdf

Lee, J. S. (2019a). Quantity and diversity of informal learning of English. *Language Learning & Technology*, *23*(1), 114–126.

Lee, J. S. (2019b). Informal digital learning of English and second language vocabulary outcomes: Can quantity conquer quality? *British Journal of Educational Technology*, *50*(2), 767–778. doi:10.1111/bjet.12599

Leis, A., Tohei, A., & Cooke, S. D. (2015). Smartphone assisted language learning autonomy. *International Journal of Computer-Assisted Language Learning and Teaching*, *5*(3), 75–88. doi:10.4018/IJCALLT.2015070105

Lewis, M. S. (2016). *Massively multiplayer online roleplaying gaming: Motivation to play, player typologies, and addiction* (Doctoral dissertation). The Ohio State University.

Liu, G. Z., Hwang, G. J., Kuo, Y. L., & Lee, C. Y. (2014). Designing dynamic English: A creative reading system in a context-aware fitness centre using a smart phone and QR codes. *Digital Creativity*, *25*(2), 169–186. doi:10.1080/14626268.2013.836110

Liu, X. Q., Zheng, D. P., & Chen, Y. S. (2019). Latent classes of smartphone dictionary users among Chinese EFL learners: A mixed-method inquiry into motivation for mobile assisted language learning. *International Journal of Lexicography*, *32*(1), 68–91.

Looi, C. K., So, H. J., Chen, W., Zhang, B., Wong, L. H., & Seow, P. (2012). Seamless learning. In N. M. Seel (Ed.), *Encyclopedia of the Sciences of Learning* (pp. 2975–2979). Boston, MA: Springer. doi:10.1007/978-1-4419-1428-6_251

Mason, A., & Zhang, W. (2017). An exploration of the use of mobile applications to support the learning of Chinese characters employed by students of Chinese as a foreign language. In Q. Kan & S. Bax (Eds.), Beyond the language classroom: Researching MOOCs and other innovations (pp. 99–112). Dublin, Ireland: Research-publishing.net. doi:10.14705/rpnet.2017.mooc2016.674

Masood, M., & Halimi, N. (2019). Gamified MALL: Are Game Elements Key to Learning Mandarin? *Advanced Science Letters*, *25*(1), 83–87. doi:10.1166/asl.2019.13192

Mayer, R. E. (2009). *Multi-media learning* (2nd ed.). Cambridge, UK: Cambridge University Press. doi:10.1017/CBO9780511811678

Morgana, V. (2018). *The IPAD and the Development of Speaking and Writing in the Secondary EFL Classroom* (EdD thesis). The Open University.

Munday, P. (2016). The case for using DUOLINGO as part of the language classroom experience. *RIED: Revista Iberoamericana de Educación a Distancia*, *19*(1), 83–101.

Myers, M. J. (2000). Voice recognition software and a hand-held translation machine for second-language learning. *Computer Assisted Language Learning*, *13*(1), 29–41. doi:10.1076/0958-8221(200002)13:1;1-K;FT029

Naderi, N., & Akrami, A. (2018). EFL learners' reading comprehension development through MALL: Telegram groups in focus. *International Journal of Instruction*, *11*(2), 339–350. doi:10.12973/iji.2018.11223a

Orehovački, T., Bubaš, G., & Kovačić, A. (2012). Taxonomy of Web 2.0 applications with educational potential. In *Transformation in teaching: Social media strategies in higher education* (pp. 43–72). Informing Science Press.

Osman, M., & Chung, P. W. H. (2011). Language learning using texting and wiki: a Malaysian context. In *e-CASE & e-Tech International Conference* (pp. 1888-1903). Knowledge Association.

Peters, K. (2007). M-learning: Positioning Educators for a Mobile, Connected Future. *The International Review of Research in Open & Distributed Learning*, *8*(2), 1–17. doi:10.19173/irrodl.v8i2.350

Rachels, J. R., & Rockinson-Szapkiw, A. J. (2017). The effects of a mobile gamification app on elementary students' Spanish achievement and self-efficacy. *Computer Assisted Language Learning*, *31*(1-2), 72–89. doi:10.1080/09588221.2017.1382536

Rankin, Y. A., Gold, R., & Gooch, B. (2006, September). 3D role-playing games as language learning tools. In Eurographics (Education Papers) (pp. 33-38). Academic Press.

Reigeluth, C. M. (2016). Instructional Theory and Technology for the New Paradigm of Education. *Revista De Educación a Distancia*, (50). Retrieved July 20, 2019 from https://revistas.um.es/red/article/view/270781

Reiser, R. A. (2012). A history of instructional design and technology. In R. A. Reiser & J. V. Dempsey (Eds.), *Trends and issues in instructional design and technology* (pp. 17–24). Boston, MA: Pearson Education.

Rosell-Aguilar, F. (2017). State of the app: A taxonomy and framework for evaluating language learning mobile applications. *CALICO Journal*, *34*(2), 243–258.

Rosell-Aguilar, F. (2018). Autonomous language learning through a mobile application: A user evaluation of the busuu app. *Computer Assisted Language Learning*, 1–28.

Schrum, L., & Levin, B. B. (2009). *Leading 21st century schools: Harnessing technology for engagement and achievement*. Thousand Oaks, CA: Carwin.

Schugurensky, D. (2000). *The forms of informal learning: Towards a conceptualization of the field* (Unpublished doctoral dissertation). University of Toronto, Toronto, Canada.

Şendurur, E., Efendioğlu, E., Çalışkan, N. Y., Boldbaatar, N., Kandin, E., & Namazli, S. (2017). The m-learning experience of language learners in informal settings. In I. A. Sanchez & P. Isaias (Eds.), *Proceedings of the 13th International Conference Mobile Learning, M-Learning 2017* (pp. 119-123). Budapest, Hungary: Academic Press.

Seo, M., Kim, J. H., & David, P. (2015). Always connected or always distracted? ADHD symptoms and social assurance explain problematic use of mobile phone and multicommunicating. *Journal of Computer-Mediated Communication*, *20*(6), 667–681. doi:10.1111/jcc4.12140

Shadiev, R., Hwang, W. Y., & Liu, T.-Y. (2018). Investigating the effectiveness of a learning activity supported by a mobile multimedia learning system to enhance autonomous EFL learning in authentic contexts. *Educational Technology Research and Development*, *66*(4), 893–912. doi:10.100711423-018-9590-1

Sharples, M., Taylor, J., & Vavoula, G. (2010). A theory of learning for the mobile age. In *Medienbildung in neuen Kulturräumen* (pp. 87–99). VS Verlag für Sozialwissenschaften. doi:10.1007/978-3-531-92133-4_6

Shi, Z., Luo, G., & He, L. (2017). Mobile-assisted language learning using WeChat instant messaging. *International Journal of Emerging Technologies in Learning*, *12*(2), 16–26. doi:10.3991/ijet.v12i02.6681

Shih, R. C. (2017). The effect of English for specific purposes (ESP) learning-language lab versus mobile-assisted learning. *International Journal of Distance Education Technologies*, *15*(3), 15–30. doi:10.4018/IJDET.2017070102

Steel, C. H., & Levy, M. (2013). Language students and their technologies: Charting the evolution 2006-2011. *ReCALL*, *25*(3), 306–320. doi:10.1017/S0958344013000128

Stevenson, M. P., & Liu, M. (2013). Learning a language with Web 2.0: Exploring the use of social networking features of foreign language learning websites. *CALICO Journal*, *27*(2), 233–259. doi:10.11139/cj.27.2.233-259

Sun, Z., Lin, C.-H., You, J., Shen, H.-J., Qi, S., & Lou, L. (2017). Improving the English-speaking skills of young learners through mobile social networking. *CALL*, *30*(3-4), 304–324. doi:10.1080/09588221.2017.1308384

Terantino, J. (2016). Examining the effects of independent MALL on vocabulary recall and listening comprehension: An exploratory case study of preschool children. *CALICO Journal*, *33*(2), 260–277.

Titova, S., & Samoylenko, O. (2017). Enquiry-based approach in mobile-supported classroom to develop language skills. *Journal of Language and Education*, *3*(3), 39–49. doi:10.17323/2411-7390-2017-3-3-39-49

UNESCO. (2011). *UNESCO mobile learning week report*. Paris: UNESCO. Retrieved March 18, 2019, from http://www.unesco.org/fileadmin/MULTIMEDIA/HQ/ED/ICT/pdf/UNESCO%20MLW%20report%20final%2019jan.pdf

Vurdien, R. (2017). Mobile assisted vocabulary acquisition and wikis to enhance writing skills. *International Journal of Computer-Assisted Language Learning and Teaching*, *7*(2), 1–21. doi:10.4018/IJCALLT.2017040101

Warschauer, M. (2000). The death of cyberspace and the rebirth of CALL. *English Teachers'. Journal*, *53*(1), 61–67.

What is 5G? (n.d.). Retrieved July 3, from https://www.verizon.com/about/our-company/5g/what-5g

Wrigglesworth, J. (2019). Using smartphones to extend interaction beyond the EFL classroom. *Computer Assisted Language Learning*, 1–22. doi:10.1080/09588221.2019.1569067

Zaki, A. A., & Yunus, M. M. (2015). Potential of Mobile Learning in Teaching of ESL Academic Writing. *English Language Teaching*, *8*(6), 11–19.

Zhang, S. (2016). Mobile English learning: An empirical study on an APP, English fun dubbing. *International Journal of Emerging Technologies in Learning*, *11*(12), 4–8. doi:10.3991/ijet.v11i12.6314

Zou, B., & Li, J. (2015). Exploring mobile apps for English language teaching and learning. In F. Helm, L. Bradley, M. Guarda, & S. Thouësny (Eds.), Critical CALL – proceedings of the 2015 EUROCALL conference, Padova, Italy (pp. 564–568). Dublin, Ireland: Research-publishing.net. doi:10.14705/rpnet.2015.000394

ADDITIONAL READING

Aslan, S., & Reigeluth, C. M. (2013). Educational technologists: Leading change for a new paradigm of education. *TechTrends*, *57*(5), 18–24. doi:10.100711528-013-0687-4

Bennett, E. E. (2012). A four-part model of informal learning: Extending Schugurensky's conceptual model. *Adult Education Conference.* Retrieved April 23, 2019, from http://newprairiepress.org/aerc/2012/papers/3

Hwang, G. J., & Fu, Q. K. (2019). Trends in the research design and application of mobile language learning: A review of 2007-2016 publications in selected SSCI journals. *Interactive Learning Environments*, *27*(4), 567–581. doi:10.1080/10494820.2018.1486861

Orehovački, T., Bubaš, G., & Kovačić, A. (2012). Taxonomy of Web 2.0 applications with educational potential. In *Transformation in teaching: Social media strategies in higher education* (pp. 43–72). California: Informing Science Press.

Reigeluth, C. M. (1992). The imperative for systemic change. *Educational Technology*, *32*(11), 9–13.

Reigeluth, C. M. (1996). A new paradigm of ISD? *Educational Technology-Saddle Brook NJ*, *36*, 13–20.

Reigeluth, C. M., Beatty, B. J., & Myers, R. D. (Eds.). (2016). *Instructional-design theories and models, Volume IV: The learner-centered paradigm of education.* Routledge. doi:10.4324/9781315795478

Reiser, R. A. (2012). A history of instructional design and technology. In R. A. Reiser & J. V. Dempsey (Eds.), *Trends and issues in instructional design and technology* (pp. 17–24). Boston, MA: Pearson Education.

KEY TERMS AND DEFINITIONS

Gamification: It is an approach including such game elements as badges to increase motivation of learners.

Informal Learning: It is a learning type taking place outside the boundaries of formal learning.

M-Learning: It is a form of digital learning happening via mobile devices.

MALL: It is a language learning process taking place with the help of mobile applications.

Mobile Application: It is a kind of software including any content/service/interaction/aim and compatible with mobile devices.

User Experience (UX): It is the whole of experiences users engage in tools, software, device, system, service, etc.

Web 2.0: The second generation of Web converting consumers into producers.

Chapter 3
Impediments to Gauging ICT-Based Informal English Learning Outcomes

Taha Ahmed Hezam
https://orcid.org/0000-0001-5779-0574
University of Bisha, Saudi Arabia

ABSTRACT

This chapter is a product of the impediments to measuring ICT-based informal learning in the Saudi EFL context. It unravels difficulties of measuring the outcomes of informal English learning with a particular reference to informal technology-based activities. It succinctly discusses why it is difficult to design, develop, implement, and sustain a comprehensive approach to assessing ICT-based informal English learning outcomes. The major problems surfaced from this discussion included absence of theoretical background of this evolving mode of learning, unawareness of recognizing linguistic, and inability to control the informal ICT-based activities for they are arbitrary and difficult to expect and measure by ordinary tests. Thus, there is a need to adopt alternative assessment tools such as portfolio, progress tests, and self-reporting within a new paradigm shift towards learner-centeredness.

INTRODUCTION

The rapid evolution of digital technology, also known as ICT, with its various and ever-growing tools and applications brought about different learning modes. The recent learning mode is independent learning through a range of ICTs that promote language learning in virtual spaces – through wired and wireless devices (Guth,

DOI: 10.4018/978-1-7998-2116-8.ch003

Copyright © 2020, IGI Global. Copying or distributing in print or electronic forms without written permission of IGI Global is prohibited.

2009; Jung, 2006; Sockett, 2014; Toffoli & Sockett, 2015; Zanetti, 2017). Prior research shows that ICTs generally help language learners to elevate their linguistic abilities. A good number of earlier studies examined the impact of ICT tools and applications on target language skills (e.g. Al-Kadi, 2018; Blake, 2008; Gonzalez & Louis, 2013; Kenning, 2007; Sharples, Arnedillo-Sanchez, Milrad & Vavoula, 2009). A plethora of research has scrutinized the merits of certain types of digital technologies in certain language aspects. For instance, Hegelheimer and O'Bryan (2009), Travis and Joseph (2009) suggested podcasting-based projects to improve oral skills. Alm (2015) and Vivian (2011) studied informal exposure to English through Facebook. Nevertheless, using technology in second language (L2) situations along with measuring its outcomes are not always straightforward. Gauging the linguistic gains stemming from free and spontaneous uses of digital technologies is an issue that warrants further research to come up with conclusive findings. There is a worldwide debate on how to measure informal language uptake that mostly occurs in informal settings facilitated by various appliances. Investigating such a topic provides implications for practical procedures to include informal practices as additional benefits to the existing formal learning. It also charts new venues for research in the Arab context where English language teaching goes through serious problems related to lack of authentic exposure, teacher training, and paradigm shifts from the traditional teaching practices to recent trends that lay heavy emphasis on learners and learning.

Recent approaches to L2 learning recognized learning that cross the borders of the classroom. Thanks to modern technology learning opportunities are handy and accessible everywhere. Alan Rogers, one of the premier authors who addressed informal learning, believes that formal and informal learning are inseparable. Rogers (2014) mooted formal and informal learning as continua – ranging from the very informal (incidental and accidental) learning through self-directed learning opportunities (self-planned) and non-formal (planned by others, not learners) to the very formal (scheduled) learning. Inspired by Rogers' continuum, Bagdonaite-Stelmokiene and Žydžiūnaitė (2016) viewed learning as a range of individual-centered learning opportunities to tacit learning (formal learning) – some of these learning activities are purposeful and some are chancy: self-regulated, self-directed, self-managed, experiential, incidental/accidental learning. Despite gaining acceptance, informal learning has not been fully researched. It remains a complex area of investigation and difficult to control its variables. One of the major problems is the difficulty of measuring outcomes of this type of learning. This provides a rationale for a close examination of this issue.

The fact that unstructured knowledge is difficult to measure, this chapter highlights salient reasons that make measuring informal learning an easy task. Putting the issue in the Saudi EFL context, this work delves into difficulties associated with informal English learning, making use of available research findings, insights from other researchers (Golding, Brown & Foley, 2009; Hager, 2006; Smith & Clayton, 2009; Werquin, 2010) as well as reflections from the author's teaching EFL context. The Kingdom of Saudi Arabia, being a rich country, has ample ICT facilities and resources. Technology is obviously used in formal education across the country. However, the evaluation system of EFL learners' performance goes under the sway of exam-oriented approach; learners are evaluated mainly through formal exams and term papers. Informal ICT-based activities are not always considered as an 'added value' to the existing educational outcomes. That is to say, learner-initiated practices via technology go mostly unnoticed by formal curriculum designers and educators. Hence, this chapter focuses on this area of research to find out reasons and justifications for the negligence of informal practices by pedagogues and decision-makers. It discusses difficulties of designing and implementing a comprehensive approach to assessing informal English learning ensued from ICT-based activities in the Saudi context as a working example of EFL scenario in the Arab countries in general and the Gulf in particular.

ASSESSMENT OF DIGITAL LANGUAGE LEARNING

Knowledge is encompassed now not only in a written form in books to be tackled and assessed but also in myriad electronic mediums which are hard to shape into measurable means. This has given way for learning without books, without teachers, and even without schools and universities. This form of learning was first disregarded. When electronic learning gained worldwide recognition, informal learning has been incorporated - skeptically though - within the concept of digital learning (Kaminskienė & Stasiūnaitienė, 2013). Although digital language learning has been evidenced as a new learning ecology (Lai, Khaddage, & Knezek, 2013), it has been described as murky and patchy as it lacks systematic evaluation (Al-Kadi, 2017; Derbel, 2017). Imbued with digital technologies, the concept of informal learning is popularized now as one form of digital learning. Informal language learning outside educational spheres, which has become acceptable in the worldwide context, gave way to new challenges, chiefly challenges measuring its outcomes and its relevance to the mainstream of learning.

Measuring Informal Learning Outcomes

Learning is now generally classified into formal vs. informal learning. The outcomes of the former are manageable and testable while that of the latter are difficult to weigh ordinarily. Informal learning generally refers to any learning opportunity which happens behind formal education (Hager, 2006; OECD, 2010; Werquin, 2010). To limit the scope of discussion in this chapter, the major focus is on informal learning that is instigated by digital ICTs. Actually, the emergence of the concept of informal learning brings to the foreground differences between formal and informal assessments. Unlike formal language assessment, informal assessment checks language comprehension and retention without scoring. Perhaps, this is quite similar to formative assessment. Although the classroom assessment is occasionally dressed in formative clothing it is overwhelmingly summative. The feedback of this assessment type is often summative, e.g., very good, excellent, good job. This summative assessment does not provide an accurate assessment of informal learning; it better suits the structured learning in the classroom.

Assessing language development outside the classroom or in non-traditional learning environments including those chances in which technological innovations are evident is an uneasy task. Gauging the effects of ICT tools on learners in informal situations has been a desirable endeavor that researchers have not fully and satisfactorily addressed. Language intake resulting from unplanned education falls within interdisciplinary areas such as language proficiency, technology-enhanced learning and the like. Earlier researchers such as Falk and Dierking (2000) suggested evaluating informal learning as a method of improving the process of learning and the ability of the institution to teach capably. The authors contend that the difficulty in evaluating informal learning is not due to lack of evidence but lack of appropriate tools. The diversity of informal learning venues and channels complicates measuring those informal outcomes resulting from informal unpredictable activities. Technology-based activities are particularly even more complicated to assess (Bonk & Lee, 2017; Hasan, Rashid, Yunus, Mohamed, & Zulkifli, 2016) because the rapidly emerging ICT innovations create a problem as to what ICT to assess and how (Knezek, Lai, Khaddage, & Baker, 2011). Today's learners have hands on a wide range of digital tools, applications, and mobile devices (wired and wireless). The WebQuests, simulations, educational games, SMS, emails, cell phones, electronic message boards, video conferences, moblogs are common examples of handy ICTs. It is difficult to study the complexity of L2 development via complex technological environments.

Obstructions of Measuring Informal Learning

There is no shortage of literature on common uses of such technologies in L2 situations (Al-kadi, 2018; Alm, 2015; Blake, 2008; Jung, 2006; Kenning, 2007; Travis & Joseph, 2009). Nevertheless, measuring the benefits ensuing from this electronic intake is underrepresented in the body of literature, and this chapter addresses it with relevance to the status quo of EFL in the Saudi context as an example of the Arab World countries. The chapter builds on the results of similar studies (e.g. Al-kadi, 2018; Falsgraf, 2009; Khaddage, Müller & Flintoff, 2016; Zanetti, 2017) and it brings to focus similarities and dissimilarities between those challenges in the context at hand. Falsgraf (2009) provided laudable obstacles of a healthy assessment of less structured learning in an open learning environment that can effectively motivate students to continue learning in a non-traditional learning environment. One of these problems, the authors highlighted, is a lack of training in assessment practices. A lack of experience in this regard inevitably means inadequate knowledge and background to take convincing decisions about instruments to measure actual learning including chances which result from informal uses of technology. Moreover, classroom assessment, which is summative in nature, generally encourages superficial and rote learning (Falsgraf, 2009). The rarity of formative assessment was another important factor that impedes informal assessment. One more vital factor is political pressure for accountability ensures that "external, high-stakes tests will be required to document school and student accountability" (Falsgraf, 2009, pp. 495-496). As such, measuring and analyzing informal learning in the digital age imply considering learning that happens outside the traditional classroom. Falsgraf's discussion of assessment difficulties explains only part of the issue.

Measuring the efficiency of ICT-based learning is even more difficult to gauge. In the literature, there is no consensus on what to achieve from an informal activity. There is similarly a disagreement on the methods, techniques, and tools of measuring such unplanned and unpredictable learning outcomes. Khaddage et al. (2016) sketched salient challenges of informal learning (outlined in Figure 1). These challenges include lack of a clear definition, technology and research challenges, and policy-making.

In addition to these challenges, Hasan et al. (2016) underscored worries resulting from the insecurity of online information sharing. These worries include "online bullying, the disclosure of personal information, access to inappropriate content, addiction, and miscommunication" (p.42.). This chapter builds on such problems and extends discussion within a wider overview of informal learning paradigm.

Figure 1. Key challenges of informal learning (adapted from Khaddage et al., 2016)

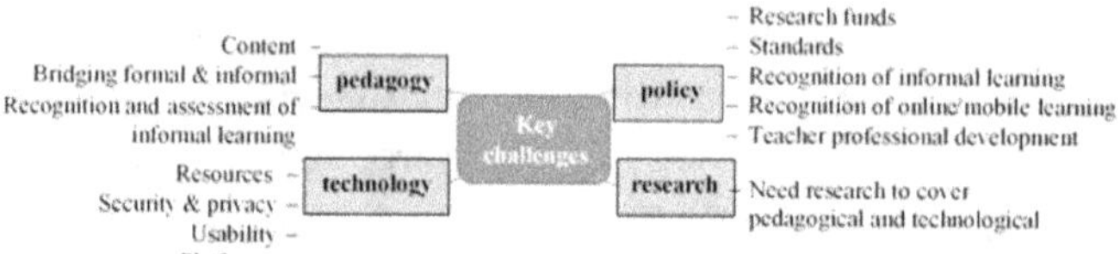

Lack of Recognition

Quite recently, informal learning is high on the agenda of contemporary educational research. It is often disused in contrast to formal learning. Both are likened to an iceberg in which we see the tip of it and the larger part is hidden (Rogers, 2014) and usually difficult to measure its size. The tip of it, which is visible and measurable, stands for the formal learning which takes place in formal institutions and is subject to official assessment. Various tests measure the proficiency of learners. Examples include the widely-used TOEFL, TOEIC, and IELTS, in addition to formal tests that every institute sets up for their learners. Nonetheless, depending on formal tests per se is insufficient to provide true indicators of real levels of the learners who undertake these tests (Al-Kadi, 2018; Jenkins, 2006). Learners nowadays have accessible mediums of learning in the absence of their teachers, parents or any other formal agencies. Such informal learning opportunities make the other part of the iceberg difficult to estimate or even design tools to measure its magnitude.

Viewing the problem from another angle, Clarke (2005, in Halliday–Wynes, & Beddie, 2009) noted that difficulty of measuring informal learning lies in the focus of assessment. Assessing informal learning requires a shift from measuring 'outcomes' to assessing 'learning conditions'. That is, assessing informal learning means considering opportunities for informal learning to ensue. Discussing challenges to validating informal learning in Europe, O'connor (2015) reflected on an ambitious aim regarding provisions of enabling individuals to have their knowledge, skills, and competences acquired via informal learning as well as obtain qualifications based on validated non-formal and informal learning experiences. Qualifications obtained online are not recognized by formal agencies (Hager, 2006; Werquin, 2010). Until now, few countries including South Africa and Ireland award credential based on knowledge gained via informal learning, while the majority of countries tend to deny recognition of online certification. That is, denying online qualifications probably discourages people to invest technological innovations for learning informally (Khaddage et al., 2016, p. 18). This requires a worldwide formal policy framework for this type of learning to be recognized (Knezek et al., 2011). In the same landscape of research, Werquin (2010) asserted that informal learning normally comes from the labor market.

Putting the issue in the educational context, informal learning arises from the need to formalize this mode of learning for it has become acceptable in the worldwide context. Nevertheless, some educators feel that formalizing all types of learning may lead to a narrower conception of learning. Recognizing informal learning could bring about constraints that might again formalize the learning, such as the requirement of formal settings (Khaddage et al., 2016).

In the Saudi context, the dominance of formal learning and its vicissitudes are evident. In spite of a big deal of reforms that the Saudi educational system has bestowed and despite significant strides in learners' assessments, there is still a common belief that formal examinations are vital to assign passing or failing marks to the learners. Notwithstanding considerable time invested on the Internet and digital devices that are assumingly good tools of learning, EFL learners in the Saudi higher educational institutions sit for classroom-based exams to measure their achievements in a given course. Online programs are not equally valued as formal educational programs. That is to say, a considerable portion of their learning, which is informal, remains unmeasured because the current teaching approach does not recognize those informal chances of learning. In such an exam-oriented learning environment, teachers prepare learners for the examinations, leaving a little space for self-assessment, self-driven learning, and self-evaluation. Currently, learners' performance, proficiency, and progress undergo formal assessment tools such formats as multiple choice questions and matching activities in the form of quizzes, midterms, and final examinations - all of which are geared towards vocabulary and grammar. Thereby, students' learning is based on rote learning and memorization. Lack of linkage of the informal activities with the formal assessing measurements leaves a quite big portion of real learning unmeasured. This creates an inconsistency between the aims, current assessments and real practices which are mostly informal and thus ignored when it comes to assessing the overall performance of learners.

Lack of Method

Another salient challenge appertains to methodology. Collecting data on informal language learning is not easy. For instance, it is difficult to look into the content of users' devices because most of the data on them are rarely aggregated and published; they are for personal uses. Accessing the contents, in some cases, requires complicated consent approval procedures. This limits monitoring and evaluating what the learners are actually engaged in informal settings (O'connor, 2015). Previous studies collected data through attitudinal measurements, which believingly leave a grey area unexplored (i.e. what learners actually do when they learn informally). Bonk and Lee (2017) discussed the challenges of success of informal learning through a survey study. The study revealed the following common obstacles as perceived by the participants. While the list of challenges that Bonk and Lee have come up with may be working in some contexts, it is not generalizable to the Saudi context. For instance, the Saudi kingdom, being a rich country, has no financial difficulties to obtain ICT devices. Again, lack of time to most of the Saudi citizens is not a big deal; they have sufficient time for learning on their own: leisure, pastimes, hobbies, networking, etc.

A lack of assessment processes, tools, and models is an important factor that complicates informal assessments. Any learning assessment is connected to corresponding documentation of competency development (Khaddage et al., 2016). The foremost challenge is unavailability of adequate and watertight diagnostic tools to measure language proficiency that stems from informal learning opportunities. Again, there is no agreement on defining language proficiency in light of technology advancements in the 21st century. There is also a lack of test that measures a broader range of learners' communicative abilities. One more thing, the upsurge of English varieties today as well as the current status of English being the lingua franca of the world, intensify difficulty of selecting assessment tools to determine one's proficiency (Al-Kadi, 2018). The world Englishes make diversity of selection and it is difficult, if not impossible, to find a concise measurement to test the overall proficiency of an English user. Let alone, the source of this proficiency, be it the result of formal learning or informal language uptake from spontaneous uses of digital technologies.

Weersing (2018) argued that the difficulty of measuring informal learning is due to its incidental nature. Learning that occurs through digital technology such as posting on Facebook, twitting on twitter and YouTube videos are difficult to predict their outcomes. Such electronic content is unplanned to be educational, and have no obvious measurable objectives. In contrast, formal learning is easy to measure and evaluate for it is sequential, linear, or compliance training delivered in the classroom, the virtual classroom, or web-based eLearning. Informal learning by its nature is complicated, amorphous, cloud-like, and arbitrary (Hajar, 2006; Khaddage et al., 2016). The outcomes of most learning activities that happen informally and are not filtered through the formal learning mediums are "imprecise and unmeasurable" (Golding, Brown & Foley, 2009, p.38). Unlike the results of formal learning are immediately testable, informal learning outcomes are difficult to gauge by ordinary tests. One more thing, informal learning outcomes generally manifest after long intervals are acquired thereby making it impossible for learners to recall where the original learning occurred. However, the skills involved had not stemmed

Figure 2. Challenges of informal learning (adapted from Bonk & Lee, 2017)

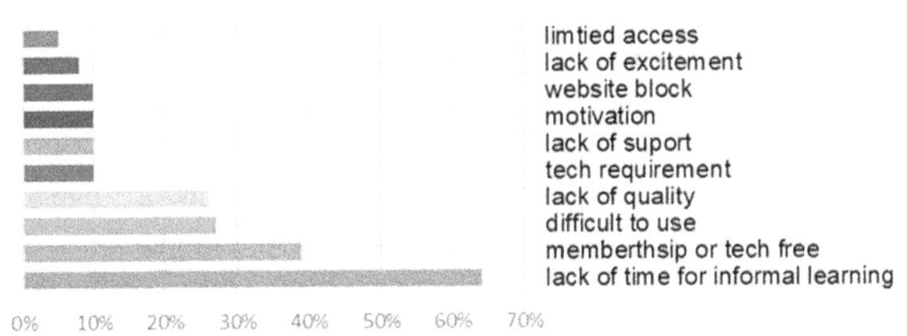

from deliberate instruction. In their study, Toffoli and Sockett (2015) found that teachers were aware that their learners were exposed to the target language through technology; yet the integration of such technologies into those teachers' courses was imperceptible. It seems that the unpredictability of outcomes of informal learning causes a sense of unease among educators.

Linguistic Concerns

Evidence from prior research showed that there are concerns related to teachers' involvement, preparation, attitude, and training. The public is unaware of the significance of exposure to the target language in the absence of teachers. Actually, the Internet with its abundances (email, google, text messaging, instant messaging, blogging, twitter, Facebook, etc.) has matured into accessible sources of language content. These electronic tools have diversified English usage and language has become expressively richer (Crystal, 2013). Nevertheless, in the context of this study, there is still a lack of awareness that these electronic means can be useful educational learning channels. Perhaps, this is because the common use of digital technology has brought about linguistic habits at the expense of declining some language skills (Al-Kadi, 2018). This has created a reaction to refrain from adopting ICT-based informal learning. Previous studies found that social media negatively affected academic vocabulary. The web-based tool such as Twitter, Facebook, and SMS texting encourage L2 learners to use an informal style of language that does not follow the normal rules of grammar. That is, the use of social media makes a person lazy to write words in full; and as a result, most students prefer using abbreviations when writing. Examples of this phenomenon might have been observed by language teachers who realize a 'dramatic decline' in the writing abilities of students due to the fad of slangy terms and odd spelling which have become common in students' assignments and written activities.

SUGGESTIONS

Building on Rogers' (2014) argument, formal learning is usually measured by formal testing, and self-planned learning is measured by tasks. Learners' personal logos, diaries, and online traces could be a good source of their language use and linguistic activities. For a better evaluation of informal learning, Carliner (2012) and Savernye (2013) suggested a multiple-method approach - a range of evaluation tools: quizzes, concept mapping, recognized acquired competencies, peer review, portfolios, performance assessment, reflective writing and media creation, rubrics, interviews, and observations, and self-reporting. These alternative assessment tools can enhance assessment of informal language learning. Bahrani (2011) examined technology-based tools including podcasts and chat for assessing the target language learners' proficiency. The author argued that such alternative tools, being valid, reliable and authentic and above all based on the students' interest, give learners a chance to use what they have learned, unlike the traditional assessments methods (paper-and-pencil assessments) which usually fail to show what learners can do with their acquired language. Bahrani suggested several activities that teachers can use with the help of technology to measure and monitor their students' achievements in language learning. By using technology, teachers can make a connection between language assessment and real-world communication.

To reiterate, there should be a competency-based approach to learning instead of the prevalent exam-oriented assessment mechanism. Assessing informal ICT-based activities should be an integral part of a flexible formal curriculum – a curriculum that promotes self-directed learning, self-monitoring, and self-assessment (Knezek et al., 2011). In a bid to establish a link between formal and informal learning, learners should be evaluated based on pre-defined competencies (skills). With mobile technologies in mind, Khaddage et al. (2016) suggested that "mobile devices provide a perfect basis for the assessment and honoring of user skills" (p.19). Lastly, yet importantly, the onus in on teachers and education deliverers to decide on what tools to employ so as to evaluate informal learning opportunities within an overall pedagogical paradigm. Adults have a fair idea of learning the first language (which to me is informal) and they have a reasonably correct measure of the competence they have achieved. But this is difficult to capture, hence a bit of monitoring by providing a protocol can be helpful.

Implications

Because of the fuzzy nature of the language construct, informal learning assessment may be equated to formative assessment including diagnostic testing and other informal assessments that classroom teachers usually use to gauge their learners' abilities in certain areas. In Kaminskienė and Stasiūnaitienė' (2013) words, 'models that emphasize assessment of non-formal and informal learning achievements are formative" (p.28). With the diversity of L2 learning opportunities nowadays, basically, via murky uses of ICTs, it is avoidable to take into account the unclear construct of the language when evaluating the overall proficiency of a language user. This is because relying on standardized proficiency tests for such purposes may be misleading; most of the available proficiency tests including TOEFL and IELTS give priority to the native speakers' model. Nonetheless, the current varieties of English in use are numerous. Being the international language with a number of local dialects all over the world and the boom of English speaking ICT mediums has weakened the significance of proficiency tests in providing true and overall measures of one's proficiency. This chapter brings to the core a plea for assessment of informal learning ensuing from ICT, apart from the traditional methods and tools of testing.

CONCLUSION

This study considers some issues of testing English learning that occurs under the influence of digital technology that language learners tend to obtain through informal uses during their presence in virtual spaces of being connected to wired and wireless devices. Guided by the focus of this chapter, the role that ICT can play in fostering language learning is only presented as a contextual factor intensifying the complexity of defining what to measure and how. Nevertheless, ICT affords tremendous opportunities for language learning and teaching. The chapter begins with a note of the assessment of digital language learning, measuring informal language learning ensued from consuming various technological innovations, and launching on impediments to assessing this type of language uptake sources including lack of method, lack of official recognition, and linguistic concerns. These snags stem mainly from a lack of adequate theory, a lack of awareness, and a flawed understanding of the concept of informal learning in the context of this study. Due to these challenges, informal learning remains hard to impose, hard to evaluate, and hard to adopt. However, I believe the situation will likely improve over time. Only when there is a formal recognition of this growing learning mode can we negotiate its expected fruits in scaffolding learners' formal education.

REFERENCES

Al-Kadi, A. (2018). *Some Aspects of ICT Uses in the Teaching of EFL at the Tertiary Level in Yemen*. Munich: Grin Publishing.

Alm, A. (2015). Facebook for informal language learning: Perspectives from tertiary language students. *The EUROCALL Review, 23*(2).

Bagdonaite-Stelmokiene, R., & Žydžiūnaitė, V. (2016). Considerations on informal learning: Different concepts and their dimensions. *Proceedings of the International Scientific Conference, 4*. 10.17770ie2016vol4.1541

Bahrani, T. (2011). Technology as an assessment tool in language learning. *International Journal of English Linguistics, 1*(2), 295–298. doi:10.5539/ijel.v1n2p295

Blake, R. (2008). *Brave new digital classroom: Technology and foreign language learning*. Washington, DC: Georgetown University Press.

Bonk, C., & Lee, M. (2017). Motivations, achievements, and challenges of self-directed informal learners in open educational environments and MOOCs. *Journal of learning for Development, 4*(1), 36-57. Retrieved from https://jl4d.org/index.php/ejl4d/article/view/195

Carliner, S. (2012). How to evaluate informal learning. American Society for Training and Development, 173-174.

Crystal, D. (2013, November 9). *The effect of new technologies on English (from the interview with David Crystal in Belgrade).* Available on YouTube: https://www.youtube.com/watch?v=qVqcoB798Is

Derbel, F. (2017). Evaluation of ICT use in language education: Why evaluate, where to look and with what means? In S. Hidri & C. Coombe (Eds.), *Evaluation in foreign language education in the Middle East and North Africa, second Language learning and teaching* (pp. 221–234). Springer International Publishing Switzerland. doi:10.1007/978-3-319-43234-2_13

Falk, J. H., & Dierking, L. D. (2000). *Learning from museums: Visitor experience and the making of meaning*. Walnut Creek, CA: AltaMira.

Falsgraf, C. (2009). The ecology of assessment. *Language Teaching, 42*(4), 491–503. doi:10.1017/S0261444808005570

Golding, B., Brown, M. & Foley, A. (2009). Informal learning: A discussion around defining and researching its breadth and importance. *Australian Journal of Adult Learning, 49*(1), 35-56. ERIC Number: EJ864431.

Gonzalez, D., & Louis, R. (2013). CALL in low-tech contexts. In M. Thomas, H. Reinders, & M. Warschauer (Eds.), *Contemporary computer-assisted language learning* (pp. 217–242). London: Bloomsbury Academic.

Guth, S. (2009). Personal learning environments for language learning. In M. Thomas (Ed.), *Handbook of research on Web 2.0 and second language learning* (pp. 451–471). Hershey, PA: Information Science References. doi:10.4018/978-1-60566-190-2.ch024

Hager, P. (2006). Recognition of informal learning: Challenges and issues. *Journal of Vocational Education and Training, 50*(4), 521–535. doi:10.1080/13636829800200070

Halliday–Wynes, S., & Beddie, F. (2009). *Informal learning: At a glance*. Adelaide: National Centre for Vocational and Educational Research.

Hasan, M., Rashid, R, Yunus, K., Mohamed, S. & Zulkifli, N. (2016). A systematic review on informal learning of English language via Facebook. *Arab World English Journal, 3,* 36-47. doi:10.2139srn.2822977

Hegelheimer & O'Bryan. (2009). Mobile technologies, podcasting and language education. In M. Thomas (Ed.), *Handbook of research on Web 2.0 and second language learning* (p. 331). Hershey, PA: Information Science References. doi:10.4018/978-1-60566-190-2.ch018

Jenkins, J. (2006). Current perspectives on teaching World Englishes and English as an international language. *TESOL Quarterly, 40,* 157–181. doi:10.2307/40264515

Jung, S. (2006). *The use of ICT in learning English as an international language* (PhD dissertation). Faculty of the Graduate School, University of Maryland.

Kaminskienė, L., & Stasiūnaitienė, E. (2013). Validity of assessment and recognition of non-formal and informal learning achievements in higher education. *Quality in Higher Education, 10,* 28–47. doi:10.7720/1822-1645.10.2

Kenning, M. (2007). *ICT and language learning: From print to the mobile phone.* Palgrave Macmillan. doi:10.1057/9780230591325

Khaddage, F., Müller, W., & Flintoff, K. (2016). Advancing mobile learning in formal and informal settings via mobile App technology: Where to from here, and how? *Journal of Educational Technology & Society, 19*(3), 16–26.

Knezek, G., Lai, K., Khaddage, F., & Baker, R. (2011). TWG 2: Student technology experiences in formal and informal learning. In *Proceedings of the ICT in Education 2011 International Summit* (pp. 1-8). Paris, France: UNESCO.

Lai, K., Khaddage, F., & Knezek, G. (2013). Blending student technology experiences in formal and informal learning. *Journal of Computer Assisted Learning, 29*(5), 414–425. doi:10.1111/jcal.12030

Levy, M. (2009). Technologies in use for second language learning. *Modern Language Journal, 93,* 769–782. doi:10.1111/j.1540-4781.2009.00972.x

O'Connor, M. (2015). Seven challenges to validating non-formal and informal learning in Europe [Web log]. Retrieved from https://ec.europa.eu/epale/en/blog/7-challenges-validating-non-formal-and-informal-learning-europe

Rogers, A. (2014). *The base of the iceberg informal learning and its impact on formal and non-formal learning.* Opladen: Barbara Budrich Publishers. doi:10.3224/84740632

Savenye, W. C. (2013). Perspectives on assessment of educational technologies for informal Learning. In *Handbook of research on educational communications and technology* (pp. 257–267). New York: Springer. doi:10.1007/978-1-4614-3185-5_21

Sharples, M., Arnedillo-Sanchez, I., Milrad, M., & Vavoula, G. (2009). Mobile learning: Small devices, big issues. In N. Balacheff, S. Ludvigsen, T. Jong, A. Lazonder, & S. Barnes (Eds.), *Technology enhanced learning: Principles and products* (pp. 233–249). Heidelberg, Germany: Springer. doi:10.1007/978-1-4020-9827-7_14

Smith, L., & Clayton, B. (2009). *Recognizing non-formal and informal learning: Participant insights and perspectives.* Adelaide: National Centre for Vocational Education Research.

Sockett, G. (2014). *The online informal learning of English.* London: Palgrave Macmillan. doi:10.1057/9781137414885

Toffoli, D., & Sockett, G. (2015). University teachers' perceptions of online informal learning of English (OILE). *Computer Assisted Language Learning, 28*(1), 7–21. doi:10.1080/09588221.2013.776970

Travis & Joseph. (2009). Improving learners' speaking skills with Podcasts. In M. Thomas (Ed.), *Handbook of research on Web 2.0 and second language learning* (pp. 313–330). Hershey, PA: Information Science References. doi:10.4018/978-1-60566-190-2.ch017

Vivian, R. (2011). University students' informal learning practices using Facebook: Help or hindrance? In R. Kwan, C., McNaught, P., Tsang, F. Wang & K. C. Li (Eds.), Enhancing learning through technology (pp. 254-267). Berlin: Springer.

Weersing, S. (2018). *How to measure informal learning.* Available at https://www.gpstrategies.com/blog/measure-informal-learning/

Werquin, P. (2010). *Recognition of non-formal and informal learning: Country practices.* OECD.

Zanetti, S. (2017). *Informal learning through ICTs: Gauging its linguistic impact on high school EFL learners and its possible application in formal learning* (MA thesis). Department of Linguistic Studies and Cultural Comparison, Università Ca' Foscari Venezia.

Section 2

Chapter 4
Learner-Initiated Language Learning Through Social Media Sites (SMSs)

Rashad Ali Ahmed
Miami University, USA

ABSTRACT

Social media sites have become an essential part of communication and interaction all over the globe. They have also offered numerous opportunities to language learners across geographic borders, paralleled by a new research interest in their potential. The present study joins this relatively new line of research as it adds data from a sample of Yemeni English language learners about their uses and perceived benefits of using social media sites in English beyond formal education. The study came up with a conclusion that Yemeni EFL learners were actively participating in social media sites and were aware of their language-related benefits. The participants reported that social media sites were helpful for building various aspects of their English proficiency but found them most useful for their writing and reading skills, expanding their vocabulary, having access to authentic materials, and communicating with English speaking friends, both native and non-native speakers. They ranked their usefulness in the following order: Facebook, WhatsApp, and Twitter.

INTRODUCTION

Over the past two decades, the Internet, particularly Web 2.0 applications, has created new channels of human communication and learning. It is now easy to access enormous and ever-expanding bodies of information online. Meanwhile,

DOI: 10.4018/978-1-7998-2116-8.ch004

Copyright © 2020, IGI Global. Copying or distributing in print or electronic forms without written permission of IGI Global is prohibited.

Social Media Sites (hereafter SMSs) such as Facebook, WeChat, Instagram, Twitter, Snapchat, and WhatsApp enable millions of people across the globe to communicate and maintain connections with their friends and relatives (Al-Kadi, 2018; Bakeer, 2018, Greenhow & Lewin, 2016; Kasuma, 2017; Vivian, 2011). Some SMS users cultivate an audience comprised of people of different ages, nationalities, religions, and cultures, who follow and interact with them by commenting on or sharing posts or even interacting virtually via face-to-face interplay. Many educational institutions, bedsides owning official pages on SMSs, are now offering learning opportunities. Students around the world spend a large portion of time present on social networking sites, performing different activities: chatting, gaming, gambling, etc. In 2008, TNS Global, a market research company, published a study on the use of digital technologies by adults from 16 industrialized nations. The study concluded that adults, on average, spend one third of their leisure time online. Most of the participants showed very high use of social networking, and over a third described social networking as 'fun' while over a quarter described it as 'interesting'.

Rationale

In the past, teachers have been generally regarded to be the primary sources of knowledge, and students depended entirely on their teachers as their exclusive providers of knowledge. With digital technology, today's language students have found new sources and resources to learn independently from parents, teachers, and other formal authorities. This trend is supported by a broader paradigm shift from teacher-dominance to learner-centered approach. Lee and McLoughlin (2008) argued that "outmoded didactic models, which place emphasis on the delivery of information by an instructor and/or from a textbook, may need to be replaced in order for student-centered learning to come to fruition" (p. 641). Now that computers and mobile technology are permeating students' lives, they spend substantial time using the Internet, seeking out new friends, playing games, watching videos, etc. (Feng, Wong, Wong, & Hossain, 2019). However, the efficiency of time that the EFL learners spend browsing social media to enhance language skills remains questionable. Feng et al. (2019) investigated how the use of Facebook and the Internet affect students' academic performance. The study found that using SMSs (e.g., Facebook) regularly for entertainment is a distraction for students and is negatively affecting their academic performance. The popularity of SMSs, specifically Facebook, demonstrates the appeal of online and virtual communities across generations, geographic locations, and cultures, but it also shows that these sites are particularly attractive to teenagers, homemakers, and students: demographics that may frequently feel isolated or lonely in their day to day lives (TNS Global, 2008).

Since TNS Global study was conducted, the use of social media has grown increasingly, especially among college students (Ha, Joa, Gabay, & Kim, 2018). As the importance of digital technology in today's world grows, so does the need to integrate these new technologies into the educational system. Thiele, Mai, and Post (2014) postulated that technology can "enhance learning by making the classroom more active and student-centered" (p. 80). Technology also offers access to authentic materials as well as opportunities for autonomous learning. Moreover, it allows language learners to communicate with native and non-native speakers from different parts of the world. Overall, the rise of modern technology among learners of English has received the attention of scholars in the field of English language teaching and substantial research exists on this issue (Al-kadi, 2018; Alsaleem, 2013; Bakeer, 2018; Chartrand, 2012; Manca & Ranieri, 2016). However, the specific language benefits of SMSs in informal settings have not been sufficiently explored. This omission serves as the primary rationale for this chapter.

LITERATURE REVIEW

Notwithstanding dozens of social networking sites, popular SMSs in the Middle East include Facebook, Twitter, WhatsApp, Instagram, WeChat, among many others. The personal utilization of these sites has become commonplace in academia. Empirical evidence in support of the beneficial effects of SMSs on students' learning and engagement remains both insufficient and inconclusive (Junco, Heiberger, & Loken, 2011). This dearth of evidence looms larger when one examines studies related to the potential benefits of SMSs for seamless informal language learning. Consequently, the literature review in this chapter is limited to the relatively small body of related studies with reference to three common SMSs: Facebook, Twitter, and WhatsApp.

Some studies have investigated the impact of social networking on education in general and on language learning in particular. The majority of these studies focused on formal language instruction (Manca & Ranieri, 2016; Kasuma, 2017; Rios & Campos, 2015). Other studies, though relatively few, found that technology has created new engaging environments for students to learn more easily and effectively outside formal institutions (Al-Kadi, 2018; Kurata, 2011; Trinder, 2017). Kurata (2011), Promnitz-Hayashi (2011), and Chartrand (2012) narrowed the focus from technology in general to the role of SMSs in language learning. The authors observed that SMSs have created new opportunities for language learners to access authentic language materials, which were difficult to find in the past. Through SMSs, learners of English can communicate with speakers from English speaking countries and non-native speakers from various countries in the world. On account of their ease to use and the emotional payoff of interacting with people that students may otherwise fail to encounter, SMSs also facilitate autonomous learning within a social environment. Similarly, Stanciu, Mihai, and Aleca (2012) conducted a study to analyze the impact of social networks on the educational process in Romanian higher education. The authors used a model for educational use of social networking adapted from Mazman and Usluel (2010). The model served to develop the usefulness of SMSs for teaching and learning in terms of six components: communication, collaboration, sharing resources, usefulness in the educational process, flexible technologies, and frequency of access. The results revealed that SMSs were very common among young people in Romania. Only 16% of the participants did not maintain SMS accounts. The study concluded that Facebook was the most common site with 87% of the participants reporting that they had Facebook accounts. Furthermore, the study found SMSs to be popular among teachers, with 56% of the participants already registered on SMSs, 44% of whom had Facebook accounts.

The following section outlines the main features of the SMSs in focus and the findings of studies that have investigated their roles in language learning. Firstly, Facebook has become one of the most frequently used social networks in the world. Statista.com reports that as of April 2019, Facebook claimed the largest number of active accounts ("2.32 billion monthly active users") and is therefore considered the most popular SMS all over the world. Facebook has many interactive features that enable users to post photos, videos, comments, add friends, and subscribe to groups. Promnitz-Hayashi (2011) observed how simple activities on Facebook helped students of a lower language proficiency class to develop various language skills. For instance, the learners who took part in the study became more comfortable while participating in online discussions, sharing opinions, and forging closer relationships with their classmates. The study reported that the incorporation of Facebook-based activities helped many of the more introverted students to be motivated to talk with their classmates. They began to express their opinions and provide extended reasoning in both face-to-face interactions and within their written classwork. In another study about Facebook as a medium of English language learning, Kabilan, Ahmad, and Jafre (2010) investigated whether students of Sains Malaysia University (USM) considered Facebook a useful learning tool that would help them learn English with greater proficiency. The researchers collected the data through a survey distributed to 300 undergraduate students at USM (81 males and 219 females). The results showed that 137 of the participants (47 male and 116 female) had Facebook accounts, whereas the remainder did not. 33% of the participants were active Facebook users, logging onto Facebook at least once a day and 52% logging on at least once a month. The study concluded that Facebook could be an effective online learning environment to facilitate learning English in terms of students' improvement of language skills and students' motivation, confidence, and attitudes towards learning English. According to the authors, "the technologies that support FB [Facebook] and features that characterize FB are able to engage students in meaningful language-based activities, even though their initial intention of joining FB is to socialize" (p. 185).

Secondly, the microblogging service Twitter enables users to text about any topic using up to 280 characters per tweet. According to Statista.com (2019), Twitter has 269 million users, and by 2020, the number is estimated to reach 275 million. Studies on Twitter have mainly examined the use of Twitter in general education or analyzed properties of the network, but few studies have focused on the role of Twitter for informal language learning. One study that investigated the usefulness of Twitter in second language learning was conducted by Borau, Ullrich, Feng, and Shen in 2009. The researchers incorporated the use of Twitter into their classrooms and made students' participation in Twitter as a part of their final grade; and this is one method of linking Twitter-based activities with formal learning. The participants were bilingual in Chinese and English and were required to provide a certain number of responses on Twitter, which yielded a total of 5574 messages. After the experiment, the students completed a questionnaire about their Twitter usage and expressed their opinions about it. The study concluded that 70% of the students found it easier to communicate in English after using Twitter, 24% had a neutral opinion, and 4.88% opposed the requirement. The informants in general had a positive view on the effects of Twitter in terms of developing English communication and cultural competence. Likewise, Junco, Heiberger, and Loken (2011) investigated whether using Twitter in the field of education had any impact on college students' engagement and grades. 125 participants between 17 to 20 years of age were divided into two groups: 70 in the experimental group and 55 in the control group. The experimental group used Twitter as part of the class, whereas the control group did not. All students answered a survey adapted from the National Survey of Student Engagement (NSSE). The results provided additional evidence that students and faculty were both highly engaged in the learning process through communication via Twitter. This suggested that using Twitter in educationally-relevant ways could increase student engagement and improve grades. Consequently, the researchers concluded that online social media could be used as an educational tool to help students reach desired college outcomes.

Thirdly, WhatsApp is an application that supports different message types, from simple text to pictures, audio files, and videos. According to WhatsApp.com (2019), more than 1 billion people in over 180 countries are active WhatsApp users at present. Alsaleem (2013) investigated the effects of WhatsApp on English writing vocabulary, word choice, and the voice or perspective of the writer. The sample included 30 undergraduate Saudi students learning English as a foreign language (EFL). The researcher used a pre-test-post-test design. The pre-test and post-test topics were similar in terms of level of difficulty and interest. The treatment involved different message writing tasks on WhatsApp. The study showed that the participants promoted their writing skills after the treatment. This improvement was particularly observed in relation to the use of vocabulary and the expression of a personal voice or perspective.

The studies presented so far share positive learning benefits from the use of SMSs for users' overall communication skills, writing and reading skills, vocabulary knowledge, and ability to express a point of view. However, there are studies that discussed negative effects of SMSs on students' learning and social life. Reporting from a state university in Turkey, Turan, Tinmaz, and Goktas's (2013) study suggested that SMSs may be a waste of time and may eventually lead to addiction to such services. The study focused particularly on Facebook. It employed qualitative research methods in the process of data collection and analysis. The researchers examined underlying reasons for the non-use of SMSs among students and found that the majority of participants did not find Facebook to be a useful learning tool. Instead, they considered it, for the most part, a waste of time, and were unable to perceive its potential benefits.

Although most of the previous empirical research related to SMSs and their use for language learning purposes mainly focused on SMSs in formal settings, this review has revealed an overall positive trend. The majority of studies have reported benefits for the users in terms of improving their communication skills, writing and reading skills, building vocabulary knowledge, and confidence in expressing their voice or point of view (Alsaleem, 2013; Borau, Ullrich, Feng, & Shen, 2009; Chartrand, 2012; Junco, Heiberger, & Loken, 2011; Kabilan, Ahmad, & Jafre, 2010; Promnitz-Hayashi, 2011). Besides these benefits, some findings have suggested more critical views of SMSs and their educational potential, pointing to the fact that they can be addictive and therefore lead to destructive behavior or are otherwise waste students' time (Greenhow & Lewin, 2016; Turan, Tinmaz, & Goktas, 2013). All SMS studies (e.g. Kabilan, Ahmad, & Jafre, 2010; Kumar, & Kumar, 2013; Rios & Campos, 2015) unanimously show that Facebook is not only the world's largest SMS but also the most popular platform for online social networking among university students. However, the bulk of prior research put emphasis on formal learning; informal language learning through these websites has been inadequately explored. More pointedly, studies on informal language learning stemming from technology-based language activities remains an obvious rarity.

Thus, this study advances research in this area by providing evidence of informal uses of SMSs to enhance formal English instruction in a context where English is not commonly used in daily situations, but serves as the medium of operating such digital technologies. This amounts to a great deal of English learning opportunities beyond the physical confinement of the classroom. The study delves into an unexplored area in the existing body of empirical literature. It outlines particular language-learning benefits gained from informal uses of SMSs, with a particular focus on Yemeni learners of English who has been underrepresented in previous studies. The study examines how this cohort of learners used SMSs informally and whether or not these sites were helpful to promote their formal English learning. The current study aims to address the following research questions:

1. What are the most useful social networking sites that help Yemeni EFL learners learn English independently from formal agencies (teachers, tutors, lab assistants, etc.)?
2. How do Yemeni EFL learners perceive the overall usefulness of social networks in relation to improving their language skills?

METHODOLOGY

This study is part of a larger quantitative and qualitative research project that thoroughly investigated the use of SMSs and gender differences impacting English learning. An online survey was sent to 100 college students at two universities in Yemen. The survey consisted of 23 questions, 20 of which were Likert scale questions and 3 were mixed-type, open-ended questions. Of the 100 potential informants, 60 responded to the survey (37 female and 23 male), yielding a response rate of 60%, which according to Nardi (2006) is a good response rate. All the participants were undergraduates, aged between 19 and 33. The overall mean age of the participants was 22 years. The rationale of selecting this body of learners was that Yemeni EFL undergraduates constituted a population of young people who tend to use online resources and networks, like the majority of young people across the World. This assumption was proven correct by the demographic data, which showed that 72% of the participants used SMSs on a daily basis, 20% used them two to three times a week, 2% used them once a week, 3.3% used them twice a month, and 3.3% rarely used them.

Reliability and Validity

The questions were designed in such a way to prevent ambiguity, redundancy, and complexity. Detailed instructions were given to the participants along with a link to the survey. Some of the questions were adapted from the research instruments of Almaghrabi (2012) and Stevenson and Liu (2010). Both reported that they had ensured the instrument's validity and reliability. Additionally, the instrument was judged by a committee of three specialists in the field of Applied Linguistics at a Midwestern University in the U.S. Their feedback and input were taken into consideration to improve the reliability and validity of the instrument. Upon data collection, Cronbach's alpha test was performed to check for the internal consistency of the survey items related to the usefulness of SMSs in participants' independent learning of English. These items were measured on a Likert scale and were thus appropriate for internal consistency analysis (George & Mallery, 2009). The results are summarized in Table 1. According to George and Mallery, a value of *alpha* = or higher than .8 shows good internal consistency. In the present study, the Cronbach *alpha* = .858 could be interpreted to mean that the 13 usefulness questions were consistently measuring participants' perceptions of the usefulness of SMSs.

Data Analysis

Descriptive statistics and frequency analyses were used to find out how frequently the participants used these networks on their own and how useful they found them for improving their English skills. The open-ended questions were subjected to content analysis to identify common themes and categories of responses. These were grouped, tabulated, and illustrated with quotes from the data.

RESULTS AND DISCUSSION

The first research question revolves around the frequency of using SMSs by Yemeni EFL learners. It aimed to examine the occurrences of SMSs uses, namely Twitter, Facebook, and WhatsApp that help students to learn English independently from their formal agencies (teachers, tutors, lab assistants, etc.). The participants were given six options to rate their SMS usage: (a) daily, (b) 2-3 times a week, (c) once a week, (d) 1-2 times a month, (e) rarely, and (f) never. The data were analyzed through frequencies and percentages of usage. The results are summarized in Table 2. As can be seen in the table, Facebook appeared to be the most commonly used SMS among the informants. Specifically, 50% of participants used Facebook on a *daily* basis, and 23.3% used it more than *2 to 3 times a week*. The second in frequency of use was WhatsApp with 36.7% of the participants using this application on a *daily* basis and 18.3% using it *2 to 3 times a week*. The third most common SMS was

Table 1. Reliability statistics

Survey section	Cronbach's Alpha	Cronbach's Alpha based on standardized items	No. of items
Usefulness Items	.858	.861	13

Twitter which was used *daily* by only 5% of the participants. Hence, Twitter was the least-used SMS among the participants, with 83.3% having *never* used it before. In addition to frequency of using SMSs in English, the participants were also asked to report how useful they found each one of the five SMSs for developing specific language skills in English. These results are presented in the following section.

Using Facebook in English

Among the three SMSs, Facebook–based on the frequency data– was the most commonly used among learners, with over 90% of the participants reporting that they had used it in English. This finding corroborated previous research that Facebook ranks first among all other SMSs (e.g. Stanciu, Mihai, & Aleca, 2012). Regarding the language benefits of using Facebook in English, participants' responses elicited the following language skills ranked in order of usefulness: writing, reading, vocabulary, grammar, listening, and speaking. The results are summarized in Table 3 where the skills are ordered from the ones that elicited the highest means of usefulness to the lowest means of usefulness. The scales 5 and 4 (*very helpful* and *helpful*) are discussed as one category '*helpful*'. The means ranged from 4.43 for writing benefits to 3.52 for speaking benefits on a five-point scale. These results supported the conclusions of Kabilan, Ahmad, and Jafre's (2010) study, which concluded that Facebook could be very useful for students in improving their language skills as well as their attitude toward English.

As is evident from Table 3, the writing skill was believed to be the most facilitated by the use of Facebook in English. Specifically, 50 out of 56 students reported that Facebook was *helpful* for improving their writing in English. Five participants reported that using Facebook was *somewhat helpful*, and only one participant believed that using Facebook was *unhelpful*. The next skill's usefulness reported was reading, as 43 participants indicated that using Facebook was *helpful*, 10 reported that it was *somewhat helpful*, and 15 reported that it was *unhelpful*. The third most useful

Table 2. Frequency of use of online social networking sites in English

SNS	N	Daily	2-3 times a week	Once a week	1-2 times a month	Rarely	Never
Facebook	60	30 (50%)	14 (23.3%)	8 (13.3%)	3 (5%)	4 (6.7%)	1 (1.7%)
WhatsApp	60	22 (36.7%)	11 (18.3%)	3 (5%)	4 (6.7%)	7 (11.7%)	13 (21.7%)
Twitter	60	3 (5%)	1 (1.7%)	0 (0%)	2 (3.3%)	4 (6.7%)	50 (83.3%)

Table 3. Participants' perceptions of usefulness of Facebook for learning English

Purpose	5 very helpful	4 helpful	3 somewhat helpful	2 unhelpful	1 very unhelpful	Ss who answered	Mean	SD
Writing	31 (51.7%)	19 (31.7%)	5 (8.3%)	1 (1.7%)	0 (0%)	56 93.4%	4.43	.735
Reading	26 (43.3%)	17 (28.3%)	10 (16.7%)	1 (1.7%)	0 (0%)	54 90%	4.26	.828
Vocabulary	26 (43.3%)	17 (28.3%)	11 (18.3%)	2 (3.3%)	0 (0%)	56 93.4%	4.20	.883
Grammar	17 (28.3%)	18 (30%)	12 (20%)	9 (15%)	0 (0%)	56 93.4%	3.77	3.77
Listening	19 (31.7%)	11 (18.3%)	10 (16.7%)	15 (25%)	0 (0%)	55 91.7	3.62	1.22
Speaking	15 (25%)	12 (20%)	16 (26.7%)	13 (21.7%)	1 (1.7%)	56 93.4%	3.52	1.12

skill was vocabulary learning as 43 students reported that Facebook was *helpful*, 11 believed it was *somewhat helpful*, and 2 thought it *unhelpful*. For grammar, 35 participants reported that using Facebook was *helpful*, 12 reported that it was *somewhat helpful*, and 9 reported that Facebook was *unhelpful*. For listening, 29 participants reported that using Facebook was *helpful*, 10 believed it was *somewhat helpful*, and 15 thought it was *unhelpful*. The least useful was reported in relation to speaking in English. Specifically, 27 students reported that Facebook was *helpful* for them to improve their speaking in English, 16 reported that it was *somewhat helpful*, and 14 believed that using Facebook was *unhelpful*. On a broader level, these findings supported Promnitz-Hayashi's (2011) findings that Facebook has become a good environment for students to take part in discussions and share their opinions, not only about daily issues, but also about issues related to their classwork.

Using WhatsApp in English

The second most frequently used application was WhatsApp, as over 76% of the participants reported using WhatsApp in English. This percentage of participants stated that they found WhatsApp useful for developing their language skills and for building their vocabulary and grammar. The informants' scores yielded the following ranking in terms of usefulness: writing, reading, vocabulary, grammar, speaking, and listening, with a mean score ranging from 4.19 for writing to 3.13 for listening. Specifically, 37 out of 47 reported that WhatsApp was *helpful* for improving their writing in English, 4 reported that using WhatsApp was *somewhat helpful*, and 6 believed that using WhatsApp was *unhelpful*.

For reading, 30 participants reported that using WhatsApp was *helpful*, 11 reported that it was *somewhat helpful* and 5 reported that it was *unhelpful*. For vocabulary, 30 students reported that WhatsApp was *helpful*, 11 believed it was *somewhat helpful*, and 6 thought it *unhelpful*. For grammar, 21 participants reported that using WhatsApp was *helpful*, 17 reported that it was *somewhat helpful*, and 9 reported that it was *unhelpful*. For speaking in English, 17 students reported that WhatsApp was *helpful*, 13 reported that it was *somewhat helpful*, and 17 believed that it was *unhelpful*. For listening, 17 participants reported that using WhatsApp was *helpful* to improve their listening in English, 9 believed it was *somewhat helpful*, and 21 thought it was *unhelpful*. Looking at Table 4, it is observed that WhatsApp helped students the most in improving their writing, with a mean score of 4.19 and a Standard Deviation of 1.11. Listening and speaking in English had the lowest mean score of 3.13. The perceived large benefits for developing learners' writing skills in English corroborated the results of Alsaleem (2013) who found that participants who used WhatsApp showed improvement in their writing skills, especially word choice.

Using Twitter in English

As mentioned in Table 2, Twitter was less popular among Yemeni students, with 83.3% of participants reporting that they had *never* used Twitter. Therefore, it ranked third in use as only 10 of the participants (16.7%) indicated that they had used Twitter in English. Table 5 summarizes the responses to the question of how *helpful* the students found Twitter in the various English skills examined. The

Table 4. Participants' perceptions of the usefulness of WhatsApp for learning English

Purpose	5 very helpful	4 helpful	3 somewhat helpful	2 unhelpful	1 very unhelpful	SS who answered	Mean	SD
Writing	26 (43.3%)	11 (18.3%)	4 (6.7%)	5 (8.3%)	1 (1.7%)	47 78.3%	4.19	1.11
Reading	18 (30%)	12 (20%)	11 (18.3%)	4 (6.7%)	1 (1.7%)	46 76.7%	3.93	1.12
Vocabulary	15 (25%)	15 (25%)	11 (18.3%)	5 (8.3%)	1 (1.7%)	47 78.3%	3.83	1.11
Grammar	11 (18.3%)	10 (16.7%)	17 (28.3%)	7 (11.7%)	2 (3.3%)	47 78.3%	3.45	1.13
Speaking	13 (21.7%)	4 (6.7%)	13 (21.7%)	15 (25%)	2 (3.3%)	47 78.3%	3.13	1.34
Listening	11 (18.3%)	6 (10%)	9 (15%)	19 (31.7%)	2 (3.3%)	47 78.3%	3.13	1.32

scales 5 and 4 (*very helpful* and *helpful*) are discussed as one category '*helpful*'. For reading, 13 participants reported that using Twitter was *helpful*, and 1 reported that it was *somewhat helpful*. For vocabulary, 12 students reported that Twitter was *helpful*, and 2 believed it was *somewhat helpful*. Out of 14, 12 students reported that Twitter was *helpful* for improving their writing in English, and 2 reported that it was *somewhat helpful*. For grammar, 7 participants reported that using Twitter helped them learn English grammar, and 7 reported that it was *somewhat helpful*. None of the participants reported that Twitter was *unhelpful* for writing, reading, vocabulary, or grammar. For listening, 7 participants reported that using Twitter was *helpful*, 5 believed it was *somewhat helpful*, and 2 thought it was *unhelpful*. For improving English speaking skills, 5 students reported that Twitter was *helpful*, 5 that it was *somewhat helpful*, and 4 that it was unhelpful. Overall, the 14 participants who reported using Twitter in English found it to be most helpful for improving their reading skills in English (Mean= 4.36, SD= .633) and to be least helpful for improving their English-speaking skills (Mean= 3.21, SD= 1.25).

These findings corroborate the observation of Stanciu, Mihai, and Aleca (2012), who found that Twitter was less popular among students from the Bucharest Academy of Economic Studies. In addition, these results agreed with Junco, Heiberger, and Loken's (2011) finding that Twitter could be used as an educational tool to help students engage and eventually lead to better learning outcomes.

The second research question is related to learners' perceptions on the overall usefulness of social networks in relation to improving their different language skills. After having examined the usefulness of each of the three SMSs separately, this

Table 5. Participants' perceptions of the usefulness of Twitter for learning English

purpose	5 very helpful	4 helpful	3 somewhat helpful	2 unhelpful	1 very unhelpful	ss who answered	Mean	SD
Reading	6 (10%)	7 (11.7%)	1 (1.7%)	0 (0%)	0 (0%)	14 23.3%	4.36	.633
Vocabulary	5 (8.3%)	7 (11.7%)	2 (3.3%)	0 (0%)	0 (0%)	14 23.3%	4.21	.699
Writing	2 (3.3%)	10 (16.7%)	2 (3.3%)	0 (0%)	0 (0%)	14 23.3%	4.00	.555
Grammar	4 (6.7%)	3 (5%)	7 (11.7%)	0 (0%)	0 (0%)	14 23.3%	3.79	.893
Listening	5 (8.3%)	2 (3.3%)	5 (8.3%)	2 (3.3%)	0 (0%)	14 23.3%	3.71	1.13
Speaking	3 (5%)	2 (3.3%)	5 (8.3%)	3 (5%)	1 (1.7%)	14 23.3%	3.21	1.25

study will examine how Yemeni students viewed the overall usefulness of these SMSs in their learning of the English language. Thirteen Likert scale questions were used to elicit participants' perceptions. The scale had three levels: 1= *unhelpful*, 2= *somewhat helpful*, and 3= *helpful*. The data were analyzed through descriptive statistics, which showed that mean scores of usefulness on all 13 items ranged from 2.17 to 2.75, indicating that the participants perceived the overall usefulness of SMSs to be from somewhat helpful to helpful. These mean scores are summarized in Table 6 in descending order, from highest to lowest mean. Table 6 clearly illustrates that the participants found the SMSs to be most helpful for improving their reading skills (Mean = 2.75). The skill of writing comes after reading in the overall usefulness with a mean score of 2.72. The third most useful aspect of SMSs was reported for learning specialized vocabulary, such as words related to sports, fashion, music, politics, etc. (Mean= 2.70). The following two useful aspects of SMSs were related to staying in touch with English-speaking friends and learning things about different cultures. They yielded the same mean values (2.67).

The participants also reported SMSs as 'useful' for making friends, both with native and non-native speakers. The two statements yielded close results; the former

Table 6. Overall usefulness of SMSs

Statements	N	Min.	Max.	M	SD
1) Improve my reading skills in **English**	60	1	3	2.75	.474
2) Improve my writing skills in **English**	60	1	3	2.72	.524
3) Learn specialized vocabulary, such as words related to 4) *sports, fashion, music, politics, etc.*	60	1	3	2.70	.530
5) Stay in touch with **English-speaking friends**	60	1	3	2.67	.542
6) Learn things about **different cultures**	60	1	3	2.67	.542
7) Make friends with **native speakers of English**	60	1	3	2.63	.610
8) Make friends with **non-native speakers of English**	60	1	3	2.62	.555
9) Learn things about **US culture**	60	1	3	2.60	.588
10) Learn everyday **English words**, including slang such as "cool", "awesome"	60	1	3	2.58	.561
11) Learn things about **British culture**	60	1	3	2.52	.596
12) Improve my listening skills in **English**	60	1	3	2.42	.787
13) Improve my **English grammar**	60	1	3	2.37	.688
14) Improve my speaking skills in **English**	60	1	3	2.17	.740

had a mean score of 2.63 and the latter of 2.62. Since the learners were exposed to British and American English only, they reported that SMSs helped them learn about the cultures of the U.S. and U.K. The mean score for learning about US culture was higher, Mean = 2.60, compared to a mean score of 2.52 for learning about British culture. The usefulness of SMSs for learning everyday English words, including slang such as "cool" or "awesome" was also rated quite highly with a mean score of 2.58. Following this was the usefulness for developing listening skills in English, which resulted in a mean score of 2.42. The highest usefulness score for developing participants' English grammar was 2.37, and the lowest was reported for developing speaking skills in English with a mean score of 2.17. Thus, SMSs were reported to be the most useful for developing participants' reading skills in English and to be the least useful for developing their speaking skills in English.

These findings support previous studies (e.g. Promnitz-Hayashi, 2011; Stanciu, Mihai, & Aleca, 2012) that have reported SMSs as helpful tools to create effective environments for students to learn a language. In a similar vein, social networking sites, by their nature, offer opportunities for practice beyond the formal classroom instruction and thereby enhances the overall language learning (Al-kadi, 2018; Greenhow & Lewin, 2016; Vivian, 2011).

The quantitative results about the usefulness of SMSs were further corroborated by the participants' narrative comments. In their answers to the question, *do you think that by using English on social networks you can develop your English skills*? The majority of the participants (83.3%) chose '*Yes*', 13.3% selected '*to some extent*', and only 3.3% chose '*No.*'

As summarized in Table 7, the narrative comments described particular benefits of SMSs. The commenters reported that although using SMSs helped them improve some language skills, these platforms were not helpful enough with other language skills. The students mentioned that when using SMSs in English, they felt they were mostly 'readers and writers' and rarely spoke or listened. Another skeptical view

Figure 1. Participants' perceptions on social networks to develop English skills

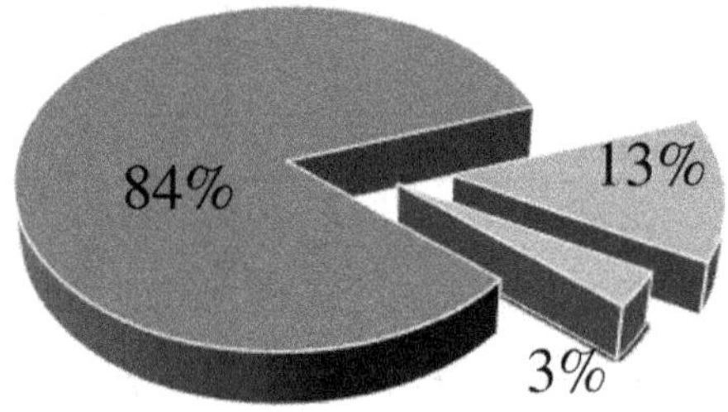

was expressed in relation to building learners' English grammar. The participants recognized that SMSs created an informal setting for communication and, therefore, grammar rules were often flouted.

The results of the current study, for the most part, support the positive view of SMS as a platform for self-determining language learning practices. Research on the potential of SMSs as language learning resources has provided controversial results. Some studies (e.g. Lee & McLoughlin, 2008; Promnitz-Hayashi, 2011; Chartrand, 2012) have found that SMSs encourage autonomous learning and offer language learners access to authentic language materials. In contrast, other studies (e.g. Turan, Tinmaz, & Goktas, 2013) have described SMSs as leading to addictive behavior with no learning benefits whatsoever. In the study at hand, these two rather opposing trends were revealed through the triangulation of the quantitative and qualitative data. However, it should be noted that the positive views were prevalent, whereas the skeptical ones were fewer and mainly concerned the usefulness of SMSs for developing grammatical accuracy and fluency in spoken skills. Moreover, the results of this study strengthen previous research findings that SMSs have been used as a means of communication and educational tools that promote autonomous learning within a social environment (Kabilan, Ahmad, & Jafre, 2010; Lee & McLoughlin, 2008; Promnitz-Hayashi, 2011; Stanciu, Mihai, & Aleca, 2012).

Regarding the current uses of SMSs, 77% of informants reported that their instructors either did not employ such platforms at all in the classroom or used them 'to some extent but not very much.' The primary rationales were of a socio-political and economic nature. Given the political climate in Yemen, which is a developing country undergoing many socio-political changes, these reasons were well-founded. Some universities still lack Internet service on campus, and classrooms are not uniformly equipped with the necessary technology. This was apparent in the comments of Participant 3 (Male) who observed that given the "circumstances in my country we barely could use social media. I just use all the theoretical type of learning. Hopefully by the very soon future, I would be able to do so."

CONCLUSION

The present study found that SMSs has been used to help students to develop their English language skills informally. In a bid to connect these informal uses of ICT tool with formal teaching, teachers need to embrace the potentials of SMSs in ways that are relevant to their students' learning. These sites serve as productive venues for exposure to authentic English. Learners present on SMSs may befriend native speakers to promote not only their linguistic abilities but also acculturation and awareness of worldwide issues. By promoting linguistic and cultural aspects, learners, belonging

Table 7. Using English on OSNSs to develop English skills

Themes	Illustrations (participants' actual quotes)	participants who contributed to a theme
Yes Category		
Improve vocabulary knowledge and language skills	*"I read many articles and messages, which help me to develop my reading skill, and providing me with many new vocabularies. It also help me to develop my skill of writing"*	40 participants out of 50 (80%)
Develop communication skills in English	*"Yes we can develop our English by being in touch with speakers of English either native speaker or non-_native speaker that we can practice English a lot"*	20 participants out of 50 (40%)
Opportunities for autonomous learning	*"It allows us more space to use English language outside the classroom through chatting with freinds via these websites either some or all."*	9 participants out of 50 (18%)
Access to information	*"Using English on social networking make me able to get the recent knowledge. And it enable me to improve all English skills."*	8 participants out of 50 (16%)
Somewhat Category		
Useful, but not for all language skills	*"to some extent cz i personally believe that not all the skills can be improved. In social networks i am just a reader and a writer. I neither speak no listen (at least for me)."*	6 participants out of 8 (75%)
Useful, but for grammar	*"I always use them for fun and chatting. When I use them, I do not care about grammar."*	2 participants out of 8 (25%)

to preservative background, can get out of their shrinks and partake actively in the worldwide arenas. Taken together, the results of the current study added supporting evidence that Yemeni college students are not an exception to the general trend of using SMSs in learning English. Based on the findings, it is implied that informal uses of SMSs can be supplementary to the usual classroom procedures, mainly in providing authentic contexts for learning, alternative assessment, and facilitating communication between students and teachers.

Despite the viable findings this study concludes, some limitations are to be acknowledged. A major limitation is related to survey-based research for the most part, where participants were asked to select from predetermined options. Another limitation was the relatively small sample size, as surveys with larger numbers of participants are more representative of the whole population than surveys with smaller samples. The sample also had unbalanced distribution of male and female participants. In addition, the study focused on the use of and language benefits of using SMSs in English, and the results should not be generalized to other languages, not to other SMSs other than Facebook, Twitter and WhatsApp. All together, these limitations might be addressed in further research studies that might build on the findings this study establishes. Further research may integrate a corpus of SMSs electronic traces on these sites and measure the relevance of that to users' linguistic abilities.

ACKNOWLEDGMENT

My deep appreciation goes to Dr. Abdu Al-Kadi for his endless support, sage advice, insightful criticism, and patient encouragement that aided the writing of this chapter in innumerable ways. My thanks also go to the students who enthusiastically participated in this study, diligently answered all questions, and provided insightful responses to the open-ended questions.

REFERENCES

Al-Kadi, A. (2018). A review of technology integration in ELT: From CALL to MALL. *Language Teaching and Educational Research*, *1*(1), 1–12.

Almaghrabi, B. K. (2012). *Saudi college students' independent language learning strategies through multimedia resources: Perceptions of benefits and implications for language* (Unpublished master's thesis). Southern Illinois University.

Alsaleem, B. I. (2013). The effect of "WhatsApp" electronic dialogue journaling on improving writing vocabulary word choice and voice of EFL undergraduate Saudi students. *Arab World English Journal*, *4*(3), 213–225.

Bakeer, A. (2018). Effects of information and communication technology and social media in developing students' writing skill: A case of Al-Quds Open University. *International Journal of Humanities and Social Science*, *8*(5). doi:10.30845/ijhss.v8n5a5

Borau, K., Ullrich, C., Feng, J., & Shen, R. (2009). Microblogging for language learning: Using Twitter to train communicative and cultural competence. *Lecture Notes in Computer Science*, *5686*, 78–87. doi:10.1007/978-3-642-03426-8_10

Chartrand, R. (2012). Social networking for language learners: Creating meaningful output with web 2.0 tools. *Knowledge Management & E-Learning: An International Journal*, *4*(1), 97–101.

Feng, S., Wong, Y. K., Wong, L. Y., & Hossain, L. (2019). The Internet and Facebook Usage on Academic Distraction of College Students. *Computers & Education, 134*, 41–49. doi:10.1016/j.compedu.2019.02.005

George, D., & Mallery, P. (2009). *SPSS for windows step by step: A simple guide reference 16.0 update*. Boston, MA: Pearson.

Global, T. N. S. (2008). *Digital world, digital life: Snapshots of our online behaviour and perspectives around the world*. Retrieved from http://www.wpp.com/~/// marketing%20insights/tns_market_research_digital_world_digital_life.pdf

Greenhow, C., & Lewin, C. (2016). Social media and education: Reconceptualizing the boundaries of formal and informal learning. *Learning, Media and Technology, 41*(1), 6–30. doi:10.1080/17439884.2015.1064954

Ha, L., Joa, C. Y., Gabay, I., & Kim, K. (2018). Does college students' social media use affect school e-mail avoidance and campus involvement? *Internet Research, 28*(1), 213–231. doi:10.1108/IntR-11-2016-0346

Junco, R. R., Heiberger, G. G., & Loken, E. E. (2011). The effect of Twitter on college student engagement and grades. *Journal of Computer Assisted Learning, 27*(2), 119–132. doi:10.1111/j.1365-2729.2010.00387.x

Kabilan, M., Ahmad, N., & Jafre, M. Z. A. (2010). Facebook: An online environment for learning of English in institutions of higher education? *Internet and Higher Education, 13*(4), 179–187. doi:10.1016/j.iheduc.2010.07.003

Kasuma, A. A. (2017). Four Characteristics of Facebook activities for English language learning: A study of Malaysian university students' needs and preferences. *Advances in Language and Literary Studies, 8*(3).

Kumar, A., & Kumar, R. (2013). Use of social networking sites (OSNSs): A study of Maharishi Dayanand University, Rohtak, India. *Library Philosophy and Practice*. Retrieved from http://digitalcommons.unl.edu/cgi/viewcontent.cgi?article=2415&context=libphilprac

Kurata, N. (2011). *Foreign language learning and use: Interaction in informal social networks*. Continuum International Publishing Group.

Manca, S., & Ranieri, M. (2016). Is Facebook still a suitable technology-enhanced learning environment? An updated critical review of the literature from 2012 to 2015. *Journal of Computer Assisted Learning, 32*(1), 503–528.

Mazman, S., & Usluel, Y. (2011). Gender differences in using social networks. *Turkish Online Journal of Educational Technology, 10*(2), 133–139.

McLoughlin, C., & Lee, M. J. W. (2008). Mapping the digital terrain: New media and social software as catalysts for pedagogical change. In *Hello! Where are you in the landscape of educational technology? Proceedings ASCILITE Melbourne 2008*. Retrieved from http://www.ascilite.org.au/conferences/melbourne08/procs/mcloughlin.pdf

Most famous social network sites. (2019). Retrieved from https://www.statista.com/statistics/272014/global-social-networks-ranked-by-number-of-users/

Nardi, P. M. (2006). *Doing survey research: A guide to quantitative methods*. Boston: Pearson.

Promnitz-Hayashi, L. (2011). A learning success story using Facebook. *Studies in Self-Access Learning Journal, 2*(4), 309–316.

Rios, J. A., & Espinoza Campos, J. L. (2015). The role of Facebook in foreign language learning. *Revista de Lenguas Modernas, 23*, 253–262.

Stanciu, A., Mihai, F., & Aleca, O. (2012). Social networking as an alternative environment for education. *Accounting & Management Information Systems, 11*(1), 56–75.

Stevenson, M. P., & Liu, M. (2010). Learning a language with web 2.0: Exploring the use of social networking features of foreign language learning websites. *CALICO Journal, 27*(2), 233–259. doi:10.11139/cj.27.2.233-259

Thiele, A. K., Mai, J. A., & Post, S. (2014). The student-centered classroom of the 21st century: Integrating web 2.0 applications and other technology to actively engage students. *Journal, Physical Therapy Education, 28*(1), 80–93. doi:10.1097/00001416-201410000-00014

Trinder, R. (2017). Informal and deliberate learning with new technology. *ELT Journal, 71*(4), 401–412. doi:10.1093/elt/ccw117

Turan, Z., Tinmaz, H., & Goktas, Y. (2013). The reasons for non-use of social networking websites by university students. *Comunicar, 21*(41), 137–145. doi:10.3916/C41-2013-13

Twitter: number of users worldwide 2020. (2019). Retrieved from https://www.statista.com/statistics/303681/twitter-users-worldwide/

Vivian, R. (2011). University students' informal learning practices using Facebook: Help or hindrance? In R. Kwan, C., McNaught, P., Tsang, F. Wang & K. C. Li (Eds.), Enhancing learning through technology (pp. 254-267). Berlin: Springer.

WhatsApp. (2019). Retrieved from https://www.whatsapp.com/about/

Chapter 5
Adult Language Learners' Informal Employment of ICT Applications and Websites to Assess Their English Skills

Ferit Kılıçkaya
https://orcid.org/0000-0002-3534-0924
Burdur Mehmet Akif Ersoy University, Turkey

ABSTRACT

This chapter attempts to investigate which ICT applications are employed by language learners to assess themselves. The study focuses mainly on 500 universities in the intensive English programs that aim to furnish students with essential language skills to pursue their studies in their subsequent departments or programs. The participants were directed to an online questionnaire in which they were asked to indicate the ICT applications that they did employ informally outside the classroom. Moreover, the participants were also asked to explain how they used these applications during the interviews. The findings of the study indicated that the participants employed websites and applications mainly for listening and grammar and that the participants had fewer options regarding speaking and writing assessment activities. Regarding the participants' choices of the applications that they could use to assess themselves, several reasons were provided that show the cost of the application and easy access and use.

DOI: 10.4018/978-1-7998-2116-8.ch005

Copyright © 2020, IGI Global. Copying or distributing in print or electronic forms without written permission of IGI Global is prohibited.

INTRODUCTION

Assessment of teaching and learning activities is crucial in any learning context including language learning and teaching. Assessment informs stakeholders of not only learners' performance but also whether objectives set for the curricula are successfully reached. More importantly, the main function of assessment is to determine and improve student learning and to act accordingly. Given that teachers' own beliefs regarding assessment and learning can affect the way assessment is conducted (Djoub, 2017, Naimi, 2018), it is almost no doubt that teachers have a crucial and indispensable role in assessment as Gareis and Grant (2015) define assessment as "... the process of using tools and techniques to collect information about student learning" (p. 2). In other words, assessment is the way teachers see their students' learning. Assessment can be conducted in a variety of tasks and for several purposes depending on the nature of learning and teaching activities and the context in which assessment is conducted.

Assessment types and tasks can be distinguished by different purposes, and uses such as placing students into appropriate classes considering their level of proficiency or determining their weaknesses and strengths so that remedial teaching and learning activities can be administered. When tests are used after the tasks and activities are conducted to determine whether learners have acquired the objectives, they are named summative assessment. However, if tests are administered during learning and teaching and the main aim is to improve instruction and to provide feedback to learners about their performance, then they are called formative assessment (Combe, Folse, & Hubley, 2007; Miller, Linn, & Gronlund, 2013). Brown and Abeywickrama (2010) define formative assessment as the type of assessment which evaluates learners during which teachers are "forming their competencies and skills with the goal of helping them to continue that growth process" (p. 7). Additionally, as teachers are required to make decisions regarding learners' performance such as assigning grades and writing performance reports, they also benefit from formal assessment and informal assessment. The former often cover traditional ways of assessment such as multiple-choice questions and fill-in-the-blank activities, while the latter includes a variety of techniques such as questioning and observing students while they are on tasks (Cunningham, 1997; Cheng & Fox, 2017). Moreover, formal assessment often include official announcement of assessment days, scores and feedback. However, informal assessment basically informs teachers of their students' performance on tasks in the classroom.

Assessment of language learners' proficiency and progress in a language is often conducted by teachers themselves locally in the classroom or through high-stakes language tests conducted nationwide as one-shot exams. However, these assessment practices appear to test only a small portion of learners' ability and knowledge

(Brown & Abeywickrama, 2010; Wright, 2015). Moreover, teachers are often held responsible for assessing learners' performance through formal assessment tools, most of which are assessment formats such as multiple-choice questions and matching activities in the form of quizzes, midterms, or final examinations. Few informal assessment practices are provided in the form of unrecorded assessment such as informing learners for strategies for reading and vocabulary difficulties and or advising on how to improve pronunciation (Wright, 2015). These examinations have positive and negative effects on teaching and learning English as a Foreign Language (EFL) in the classrooms. As quizzes mainly assess vocabulary and grammar, student learning is often based on memorization, leaving almost no room for practicing productive skills.

Prevalent uses of formal assessment and overuse of summative assessment pose several obstacles to appropriate and healthy assessment. As Falsgrap (2009) clearly voices, two of these obstacles are crucial: rarity of formative assessment and overuse of high-stakes tests. Classroom assessment is often overwhelmingly of summative nature, together with insufficient or superficial feedback based on the results. Therefore, in addition to summative assessment, formative assessment can be used as effective tools, teachers must take student-learning and needs goals into consideration (Miller, Linn, & Gronlund, 2009). As indicated by McMillan (2010), classroom environment should also be characterized by respect, tolerance, honest communication and collaboration to support formative assessment. Moreover, it is also vital that formative assessments, due to their nature, should be implemented throughout learning and teaching process for several times contrary to summative assessments conducted at the end of the units or the semester. Nonetheless, in teaching contexts where teachers are required to teach many class hours and many students in different classes, achieving effective formative assessment does not seem to be possible as the nature of formative assessments requires assessing students at different times and provide feedback to students following each assessment.

In addition to these obstacles and the belief that teachers are the ones responsible for any assessment practice in the classroom, the inconsistency between what is aimed, practiced and assessed leads to further problems in assessment (Brown & Abeywickrama, 2010). The changes introduced in high-stake exams and great importance attached to those exams force teachers to integrate the skills as tested in the examination and to feel that their major responsibility is to prepare students for examinations. Moreover, limited class time and other factors such as scoring students' papers appear to be important issues for teachers willing to conduct assessments. In fact, new technological developments have provided easy yet effective solutions to these issues.

Today's language learners are aware that assessment is no longer limited to what happens in the classroom or nation-wide exams, and teachers are not the only source of materials that allow them to be informed of their progress in the language they are learning. In informal assessment of one's own skills through the websites on the Internet and applications on mobile devices (Chinnery, 2006), learners themselves are at the very center of assessment process (Sockett, 2014). They practice language skills through interaction with these applications and websites based on their own needs, which might be considered part of learner autonomy not just from the perspective of learning but also from assessment. Due to limited nature of formal and summative classroom assessment and several factors such as delayed feedback and an insufficient number of questions possible to be included in assessment, learners' informal assessment practices with the help of technology can serve formal learning objectives such as *Quizlet* and *test-english.com*. Therefore, the current chapter aimed to investigate language learners' informal use of technology to assess their proficiency and performance. The following research questions were addressed in the study:

1. What applications and websites do university EFL learners report they use when they want to assess their skills in English?
2. Which skills do the university EFL learners assess informally by using applications and websites?
3. What are university EFL leaners' views regarding the use of applications and websites in assessing their English skills?

THEORETICAL BACKGROUND

Prevalence of digital technology, devices, and applications has paved the way for learners to discover new ways of boosting their learning through open educational resources such as online courses, websites and videos freely available on the Internet and creating their own assessment tools such as flashcards. Although it is a fact that learners use these resources mainly for learning purposes (Cheung, 2019), it is also accepted today that ICT provides learners with the opportunity to access online resources and materials and use them to assess themselves (Brown, 1997) and get immediate feedback rather than waiting to be informed about their performance. For example, learners can now respond to freely available online questions with fast and detailed feedback, moving beyond the walls of the classroom where they could continue learning and assessing themselves.

The Role of ICT Use in Learning and Self-Evaluation

The role of assessment, as it is widely acknowledged, is crucial in teaching and learning practices as they determine and measure the level of accomplishment of objectives. Similar to this role, technology is expected to "supplement and not substitute for high-quality instructional methods" (Asri, 2019, p. 346). Learning languages beyond the classroom has been investigated from various perspectives ranging from learners' motivation to participants' views on benefits (Lai & Gu, 2011; Lai, 2015). Learners might also use technological devices to check what they can do with the language and to assess what they have learned through self-assessment, which refers to the activities in which learners evaluate their own performance generally following the activities completed (Matsuno, 2009). For example, automated self-assessed quizzes allow learners to practice and review course content and related materials and enable teachers to encourage learners to practice numerous exercises based on learners' needs (Pérez-Solà, Herrera-Joancomartí & Rifà-Pous, 2018).

Regarding the impact of technology in the classroom, self-evaluation and learners' views, several studies have been conducted to determine to what extent technology and technological resources have affected learning and learners and in what ways (Wang & Smith, 2013; Golonka et al., 2014; Lai, Hu, & Lyu, 2018; Ghouali & Benmoussat, 2019; Hung, 2019). Wang and Smith (2013), for example, reported from a longitudinal study that learners' practicing grammar and vocabulary using mobile devices resulted in a positive language experience. The participants included first-year Japanese university students, and they responded to compression questions for every reading material supported with grammar quizzes and grammar point reviews. Yet, it was also noted that mobile supported learning must also be provided with a proper degree of teacher monitoring and feedback and the need for incentives. In another study, Golonka et al. (2014) analyzed over 350 studies on technology use and its affect and impact on foreign language learning and found that the results revealed moderate support for the claims that technology enhanced learners' interaction, motivation, and feedback. Lai, Hu, and Lyu (2018), on the other hand, investigated language learning with technology outside the classroom. The data were collected by interviewing 21 university language learners and 439 survey responses. The results yielded three types of technological experiences, one of which is instruction-oriented technological experience. This type of experience underscored how useful learners considered technological experience.

Investigating the impact of social media and writing skills, Ghouali and Benmoussat (2019) benefited from a case study in which 31 undergraduates and 22 English teachers at university participated. The data were collected from questionnaires given to the participants and writing tasks assigned on social media and in the classroom. The results revealed that the written task completed on social media resulted in low performance in formal writing and a failure in following the norms of the English language. Hung (2019) investigated effectiveness of repeated self-assessment of English language learners' speaking skills and their views regarding this. The participants included 97 students in a Taiwanese college, who recorded their oral responses using their mobile phones and responded to the questions to evaluate their own talks. The results showed that self-assessment conducted through repeated practice resulted in an increase in oral performance and also yielded positive perceptions.

Use of Online Websites and Quizzes

Studies have shown that learners had positive attitudes towards use of online quizzes and expressed preference over traditional exams and questions (Dumova, 2012). Moreover, several other studies have also shown the positive effects of online quizzes in various ways (Woit & Mason, 2000; DeSouza & Fleming, 2003; Bunce, VandenPlas, & Havanki, 2006; Kibble, 2007; Johnson & Kiviniemi, 2009). For example, Woit and Mason (2000) studied the effects of weekly online quizzes on students' learning in laboratory assignments of computer science courses. X-terminals and Windows/NT workstations were used to deliver quizzes through secured connects to the students. The results of the study indicated that online quizzes helped increase participants' learning and retention. DeSouza and Fleming (2003) compared the effects of traditional paper-and-pencil quizzes and online quizzes on the achievement of the participants enrolled in a Theories of Personality class. The findings revealed that the participants that took online quizzes scored significantly higher on four class exams than those who were exposed to the traditional paper-and-pencil quizzes due to several reasons such as immediacy of feedback and flexibility.

Bunce et al. (2006) analyzed the comparative effects of online quizzes provided on the platforms of WebCT and Student Response System on the achievement of the participants enrolled in a chemistry course in a small private university. The results indicated that WebCT quizzes had statistically significant positive effects on student achievement and preparation for the examinations. In another study conducted by Kibble (2007), students enrolled in a Medical Physiology class were asked to take several online quizzes before midterm and final summative examinations. The results showed that the students that responded to the questions on online quizzes performed better in the midterm and final examinations. Johnson and Kiviniemi (2009), for instance, administered quizzes on assigned readings in an introductory social psychology course to undergraduate students through a web-based testing system. The results indicated that the participants who had completed reading quizzes improved their exam and classroom performances. The study also underscored the importance of online tests as they provided students with the opportunity to move beyond the walls of the classroom where they could continue learning.

Moreover, based on university teachers' perceptions of technology, it is stated that online activities have positive effects on participants' listening comprehension, while positive effects appear to be fewer for writing and reading skills (Toffoli & Sockett, 2015) because of the nature of these skills, which requires teacher monitoring, detailed feedback and support. While it is rather easier to provide correct answers to listening comprehension questions, grammar, and vocabulary exercises, this might not be valid for writing activities, especially when students are asked to provide longer responses. However, the key requirement for positive effects for the use of technology in language learning and assessment can be stated to be use of suitable technology for the right subject and skill. That being the case, applications and websites promising the development of language skills and components can be very helpful and motivating for language learners (Sundqvist & Kerstin Sylvén, 2016). Considering this trend, this proposed chapter attempts to investigate which ICT applications and websites are employed by adult language learners to assess their English skills.

METHODOLOGY

Contexts and Participants

The study included participants who were students of preparatory EFL classes at eleven state universities. The participants were required by their universities to enroll in these preparatory classes as their level of English was not sufficient to pursue their education in their departments in which most classes are held in English. The data were generated from a survey of 500 respondents, who were all university EFL learners and were attending intensive EFL programmes in Turkey, and the semi-structured interviews with 20 participant, who were invited to express their views and experience. The participating students were all native speakers of Turkish, with an average age of 19.

Data Collection and Procedure

The questionnaire, which was developed to explore participants' employment of ICT applications and websites to assess themselves, contains the following two open-ended questions for structured responses: (a) Which websites/applications do/did you use to assess your English language skills outside the classroom? (b) Which skills do/did the university EFL learners assess using applications and websites? In order to ensure the validity and reliability of the questionnaire, expert views were obtained on the items to decide whether the questionnaire items were adequately measuring the construct intended to assess, and whether the items were sufficient to measure the intended construct in addition to the pilot test on a small sample of 40. The respondents were asked via open-ended questions to elaborate what they thought about each questionnaire item. Moreover, test-retest reliability analysis was conducted on the randomly selected 40 participants' responses on the repeated administration of the same questionnaire. The same individuals were administered the same questionnaire twice after 20 days. Pearson's product-moment correlation coefficient (Pearson's r) was calculated and determined to be +.83, indicating good reliability.

The interviews were semi-structured and included a list of questions prepared in advance. Although the interviews were limited in terms of the questions asked, the participants were also asked impromptu questions to analyze and explore the issue(s) in focus (see Appendix). The interviews focused on how the participants used the applications and websites that they provided in their responses to the questions in the questionnaire and the major benefits and limitations of using these applications and websites. The interviews also focused on which skills the participants aimed at assessing.

The investigation was carried out in two phases. In the first phase, the questionnaire was shared with all potential participants at eleven universities. The participants were directed to an online questionnaire created through Google Forms, a free and popular survey creation and publication tool, in which they were asked to indicate ICT applications and websites they used informally outside the classroom. Moreover, the participants were also asked to explain how they used these applications and websites. The second phase included the follow-up semi-structured interviews with twenty respondents. 500 of these 650 students responded to the online questionnaire. Three weeks after the data collection through the online questionnaire, a sub-sample of thirty respondents were randomly selected from the list of participants willing to participate in a semi-structured interview. These participants were interviewed in their L1 via Skype or Google hangouts. All interviews except five were recorded and transcribed later as these five students did not agree to be recorded. As an example, an excerpt from one of the interviews is provided below, together with keyword coding process used for the analysis of responses.

Participant: Most of the time I use Elllo.org because it provides short videos and most have quizzes with it.

Interviewer: How do you use this website?

Participant: I use it to watch and assess my listening. I mean I choose a suitable video and then watch it to answer the questions. I get immediate results. [coding: Immediate result]

Interviewer: So, do you see it as the major reason for using the website?

Participant: It is one of several reasons. There are different videos on different topics [coding: Variety of materials] and you can easily check yourself.

Data Analysis

The data obtained through the online questionnaire were analyzed statistically through descriptive statistics to determine the most frequently used applications and websites. The interview data, as exemplified by the excerpt above, were prone to content analysis and analyzed qualitatively. The interviews aimed at determining possible reasons why specific applications and websites were used by the participants. Therefore, several keywords emerged during the analysis, which were later used as the guidelines to group reasons for using the applications and websites. The emerging codes were checked for consistency by two EFL lecturers with experience in coding and analyzing interview data. The codes were revised based on the discussions and suggestions regarding the inconsistencies. The quotes were selected to best represent participants' views. Member checking was also utilized to enhance the trustworthiness of the qualitative analysis (Creswell, 2005), in which the participants who were interviewed checked and confirmed the transcriptions and their views. To ensure the coder reliability, Cohen's kappa coefficient was calculated between these two coders in the study, which were determined to be .75.

RESULTS

In order to answer research questions '(1) What applications and websites do university EFL learners report they used when they want to assess their skills in English?' and '(2) Which skills do university EFL learners assess using applications and websites?', the data obtained through the online questionnaire were analyzed quantitatively through descriptive statistics. Table 1 provides the applications and websites, basic description, frequency, percentages, skills targeted, and links to these applications and websites. As seen in Table 1, the participants reported a variety of applications and websites that they used to assess their own skills outside the classroom. *YouTube* appears to be the leading website (91%) that helps participants assess their listening skills, which is followed by another website, *TED*, a non-profit organization website hosting a variety of talks in various languages (74.40%). *English Language Listening Lab Online (ELLLO)* (60%) and *Randall's Cyber Listening Lab* (50%) are other two websites provided by the participants to assess their listening skills with multiple-choice quizzes. Regarding grammar, the participants reported four main applications and websites: *My Grammar Lab* (40%), *Road to Grammar*

(31%), Quizlet (30%), and finally the Grammar Aquarium (24%). Moreover, the participants reported Quizlet (30%) and Memrise (20%) to assess their vocabulary knowledge. These two websites allow teachers and students to create vocabulary flashcards and quizzes in addition to accessing materials and exercises created by other learners and teachers. While participants report various applications and websites for listening, grammar and vocabulary, few resources were reported to assess other skills such as reading, writing and speaking. For example, participants provided only one website to assess their reading: *Readlang* (20%), a website that translates words and phrases in a text. Similarly, one application was reported to be used to assess participants' writing skills, which turned out to be *Replika* (17%), a mobile application benefiting from artificial intelligence to interact with the participants in the written and spoken modes.

As for speaking, the participants claimed that they used several applications to assess their speaking: *Replika* (17%), *ESL Fast Speak* (10.8%), and *TalkEnglish* (7%). Except the first one, the other two applications provide practice by presenting dialogues in the form of conversations through which learners are asked to repeat. When some applications and websites were not voiced by more than two participants, in other words, whose frequency is less than two, they were not included in the table separately but under the name 'Other'. Some of these applications included paid services such as *Cambly*, a website where learners can find and speak to English tutors online and a reading website, *Literacynet,*

Based on the data obtained through an online questionnaire, it can be put forward that participants use applications and websites to assess mainly listening, grammar, and vocabulary. The finding showing that participants benefit more from listening activities and that this has positive effects on participants' listening comprehension corroborates that of the study conducted by Toffoli and Sockett (2015). This can be attributed to the fact that technology allows specific skills to be practiced and assessed more easily compared to other skills such as writing and speaking, which require more human elements to be used together with the technology to assess learners' essays and speech. In other words, there are few resources available for skills such as writing and speaking as these skills require more interactive oriented activities rather than one-way communication.

Although the current technological developments seem to provide promising improvements in practicing and assessing productive skills, they are either not fully developed or may cost more than individual learners can afford. This is also evident in the applications and websites used by the participants, who are more oriented towards the services provided free of charge. Moreover, when these applications and websites are analyzed in terms of the characteristics of the activities, the participants appear to have preferred the ones that provide automated self-assessment quizzes, which is in line with the findings of the study conducted by Pérez-Solà et al. (2018).

Table 1. The description of the applications/websites provided by the participants

Website/ Application	Description	Freq	%	Skill	Link
YouTube	Popular website providing various videos in numerous language	455	91	listening	https://www.youtube.com/
TED	Website of non-profit organization providing various videos including short talks on a variety of topics	372	74.40	listening/ reading	https://www.ted.com/
English Language Listening Lab Online (ELLLO)	Website providing short videos supported with online multiple-choice quizzes on various topics	300	60	listening	http://www.elllo.org/
Randall's cyber Listening Lab	Website providing short audio files with online multiple-choice quizzes	250	50	listening	https://www.esl-lab.com/
My Grammar Lab	Application used with the course book MyGrammarLab by Pearson Education, which includes multiple-choice questions	200	40	grammar	https://play.google.com/store/apps/details?id=com.pearson.grammar.advanced
Road to Grammar	Website including Flash-based grammar exercises and quizzes on a variety of English structures with explanations	155	31	grammar	http://www.roadtogrammar.com/
Quizlet	Website that allows creating flashcards and quizzes and searching exercises created by other users	150	30	vocabulary/ grammar	https://quizlet.com/
The Grammar Aquarium	Website with online quizzes and instructional materials on various grammatical structures with explanations	120	24	grammar	http://perso.wanadoo.es/autoenglish/freeexercises.htm
Memrise	Website/application allowing the use of flashcards to help retention and practice of vocabulary and providing other users' content	100	20	vocabulary	https://www.memrise.com/
Readlang	Website that translate words and phrases in a text in many languages	100	20	reading	http://readlang.com
Replika	Mobile application including an artificial intelligence character with whom you can have written and spoken communication	85	17	writing/ speaking	https://play.google.com/store/apps/details?id=ai.replika.app
ESL Fast Speak	Mobile application including audio conversations on various topics.	54	10.8	speaking	https://play.google.com/store/apps/details?id=com.eslfast.easyspeak
TalkEnglish	Mobile application aiming at improving learners' English conversation skills	35	7	speaking	https://play.google.com/store/apps/details?id=com.talkenglish.practice
Other	Other applications/websites	20	4	all skills	

To answer the third research question 'What are the university EFL leaners' views regarding the use of applications and websites in assessing their skills?', the data obtained through the semi-structured interviews held with the selected participants were subject to content analysis. The findings based on the interviews are provided under each theme and corresponding codes (Table 2). The themes and the codes were determined as: (i) Major benefits (fast scoring/result and flexibility, a variety of materials, and detailed feedback), (ii) Major disadvantages (fewer resources, cost, and reliability), (iii) Skills/components assessed (grammar, vocabulary, listening, reading, writing, and speaking).

Major benefits. Under this theme, four codes emerged: fast scoring/result, flexibility, a variety of materials, and detailed feedback. The majority of the participants (n = 26) expressed that the major benefit of using the applications and websites in assessing their skills was the fast scoring/result. This was clearly expressed in the following quote:

When I complete the quiz [a multiple-choice grammar test], the website automatically tells me my score, together with my correct and incorrect answers. I think this is great as we often wait for a week or so to learn our scores. In the classroom when we take a quiz or do an exercise, it takes often for a week or so to learn about our score and during that time we might forget our responses [Participant 5, Female].

Four participants did not agree on this with the other participants. They claimed that fast scoring or result was not always possible with all skills such as writing. The majority of the participants (n = 27) also expressed that assessment practices using the website and applications on the internet were flexible in terms of the devices and the time. One participant expressed this as follows:

Another advantage of doing online quizzes is I think the time. I can use the applications and websites almost on any device that I own and I can do it at anytime and anywhere. On the bus to the university, for example, I do vocabulary activities on my mobile phone. It takes almost an hour to go to the university and I can benefit from that time for my English [Participant 7, Male].

Several participants (n = 20) expressed that another benefit of using applications or websites in assessing their skills was the variety of materials available on the Internet. They underscored that materials on especially grammar, vocabulary and reading were abundant and they could easily find materials for these skills and language components using the search engines with several keywords. One of the participants stated tha

When I search Google for online quizzes and exercises on grammar and vocabulary, I find tons of websites I sometimes don't know which one to check. For example, when I want to do more practice on a specific language structure, say conjunctions, you can easily find quizzes in seconds. The time and the materials that could be used to assess a specific structure or vocabulary in the classroom are limited; however, you can find lots of materials on the Internet [Participant 11, Male].

Several other participants (n = 6) acknowledged that although there were many quizzes and exercises for grammar and vocabulary, there were limited opportunities for writing and speaking assessment. This is pointed out as:

I can find a lot of materials to assess my grammar and vocabulary knowledge in addition to reading. But I cannot say the same for speaking and writing. There are few resources available and most of them are paid [Participant 12, Male].

Major disadvantages. This theme included three codes: fewer resources, cost, and reliability. The majority of the participants (n = 25) shared the view that there were few resources available for speaking and writing skills, notably. Moreover, many participants also noted that several applications and websites might no longer provide their free service after some time. This is clearly expressed in the following quote:

I can find easily the quizzes, exercises, etc. for grammar and vocabulary, and to some extent reading. But it is so difficult to have websites or applications to assess our writing and speaking skills. There are some but almost all of them require paying for their services and I cannot afford them [Participant 25, Male].

Regarding reliability, one participant expressed her view as follows:

It is really saddening to see several websites go paid or stop providing their service. You start using an application and you really benefit from it. But all of a sudden it disappears or requires payment [Participant 22, Female].

As for language skills and components assessed using applications and websites, the responses of the participants confirm, to a large extent, and contribute to the results obtained through the online questionnaire. The interviews also indicated that the participants mainly used the Internet to assess their listening skills and knowledge of grammar and vocabulary to review and reinforce outside the classroom what they learned in the classroom. Two of the participants expressed this as follows:

Table 2. The themes and codes that emerged from the responses provided during the interviews

Theme	Code	Sample Response
Major benefits	fast scoring/ result	*I think the main advantage of using online resources to asses our language skills is how quickly we receive the score. I mean the results of our responses. This is valid of course for multiple-choice questions. When I finish the quiz and click on the submit or evaluate button, I quickly get the results together with the correct and incorrect answers.*
	flexibility	*I can use the websites and the applications anytime and anywhere I mean I can assess my vocabulary or grammar on my mobile device at home or on the bus. I do not have to be anywhere. I do not have to limit myself to the classroom.*
	variety of materials	*In the classroom we have limited materials or resources to assess our grammar or vocabulary or the other skills. However, on the Internet you can find different websites or applications to quickly assess your use or knowledge of English. For example, if you want to assess your knowledge of relative clauses, you can search it on Google and find tons of tests or quizzes to assess yourself.*
	detailed feedback	*Several websites provide detailed information about your performance on the test. For example, you can get a detailed analysis regarding your response, why it is correct or incorrect and you are provided with many other examples or explanation.*
Major disadvantages	fewer resources	*For several language skills and components, I have fewer resources. I cannot assess my writing for example on the Internet.*
	cost	*I know that there are some websites or applications that can assess what you have written. But they are either not free or too expensive.*
	reliability	*I find several websites very useful. I use them a lot during my studies. But all of a sudden I see that they disappeared or stopped providing the free service. Or worse, I see they disappear. I think that is the major disadvantage of these websites. You cannot be sure how long they will exist or provide free service.*
Skills/ components assessed	grammar	*After the classes when we had covered several grammatical structures, I use several websites such as 'Road to grammar' to assess my knowledge. It is great to easily find the exercises and get the correct and incorrect answers. Moreover, I am also provided with the explanation regarding the topic.*
	vocabulary	*I use several websites to practice and review the words that we learned in the classroom. I think repetition is very important and technology is great in terms of providing several opportunities to do so. For example, I benefit from vocabulary quizzes and also I create my own vocabulary quizzes to both practice and to test my knowledge.*
	listening	*Compared to the past, it is much easier to find videos on the Internet. YouTube, for example, provides great videos on any topic and at any level. I watch YouTube videos ranging from film fragments to TOEFL listening questions. In addition to this, I use ESL-lab to practice and assess my listening skill in English. The website provides audio files at various levels and I choose the ones generally related to my interest.*
	reading	*I sometimes check Literacynet to read various texts with multiple-choice questions. Moreover, I often read paragraphs similar to TOEFL or IELT questions. But, I do not prefer to read online as I find it rather difficult or distracting to read on screen.*
	writing	*It is very difficult to find websites or applications that will assess our writing. There are a few. But you have to pay a lot for their service. So, I use websites that allows practicing rewriting or paraphrasing sentences or writing short paragraphs.*
	speaking	*Speaking, I think, is one of the most challenging skills considering the limited opportunities and resources we have. Apart from the specific speaking classes, we do not have the opportunity to practice speaking. I know that some of my classmates try to find friends on Facebook or other websites to practice speaking in English. I use* Cambly *software to practice speaking but it is paid service, and not everybody can afford to pay for the service they provide.*

I mainly use several applications and websites to review and assess what I have learned in the classroom. We are given further exercises as homework to review; but on the Internet there are many and different exercises provided with explanation and quick scoring. I just search and find grammar and vocabulary activities especially. [Participant 17, Male].

Internet is great especially when it comes to listening materials. Compared to the past, we have more resources of listening materials in English. I watch YouTube videos and try to understand it. I mainly practice and assess my listening on several other websites such as ELLLO. They are great as I also respond to the questions on what I listen. I am not just limited to the classroom materials. If I feel that I need to do more, I can quickly find a lot of materials [Participant 8, Female].

As for other skills, reading, writing, and speaking, it can be stated that the participants were not content with the resources available on the Internet and that the applications and websites available for speaking and writing are scarce. One of the participants expressed that

It is almost impossible to find websites or applications that will assess our writing skills due to nature of writing. It is also the same case with speaking. Most Internet resources are based on repeating what is heard or written. Speaking is a two-way communication. So, we need to communicate. There are some websites on which you can do it but it requires payment [Participant 13, Male].

During the interviews, the participants expressed that they were provided with a variety of reading materials through their coursebooks and extra reading materials such as graded readers. Therefore, they did not raise concern regarding the assessment of their reading skills. This is evident in the following quote:

In the classroom we do a lot of reading. We are also provided with other materials such as reading worksheets and extra reading texts. In this respect, we do not need much to assess our reading skill. I sometimes check; however, it is not the case with speaking and writing skills. For writing, for example, there are just online websites that check fill-in-the blank activities or paraphrasing sentences [Participant 27, Male].

DISCUSSION AND IMPLICATIONS

Based on the data obtained through semi-structured interviews, it can be stated that the participants mainly used applications and websites to review, practice and assess skills and components such as grammar, vocabulary, and listening due to several reasons such as obtaining fast results and detailed feedback. That being the case, the participants seem to be motivated as they see their progress and development. *YouTube*, for example, although it does not aim to provide online quizzes to check listening comprehension, has been found useful by the participants as they check through the videos whether they can understand the dialogues and conversations better compared to the beginning of their studies. As indicated during the interviews, participants feel more self-confident when they see that they are performing much better through doing the activities on the Internet. As indicated by Sundqvist and Sylvén (2016), when learners use the right technology to improve and assess their skills, they are motivated to study and learn further.

The interviews also confirm the findings obtained through the online questionnaire indicating that language assessment practices outside the classroom have been found useful and beneficial by the participants, which confirms the findings underscored by Lai and Gu (2011) and Lai (2015). Moreover, based on the perceptions of the participants and their experience, practicing online quizzes not only helps the participants see their progress in their studies but also notice their strengths and weaknesses. Although this study did not aim or focus on the effect of online quizzes on learners' performance or success, this finding seems to be in line with those of several other studies (i.e. Bunce et al., 2006; Kibble, 2007; Johnson & Kiviniemi, 2009) in several ways. The participants expressed that they benefited from immediacy of feedback and flexibility as they were informed of their performance after each question and/or at the end of the exercises or quizzes and that they did the activities and quizzes whenever they wished either on their computers or their mobile devices. This, as a result, enabled the participants to take not only learning but also assessment out the walls of the classroom. Moreover, although it was stated by a few participants, online informal self-assessment practices also create for confidentiality. In other words, learners' performance and their results are not divulged and privacy helps guard their self-esteem, which, according to these participants, is a positive fact.

However, confirming the data obtained from the online questionnaire, the interviews also indicated that the participants were more in need of materials and ways of assessment of speaking and writing skills. This is clearly evident in the participants' responses as almost each participant underscored the need for more materials regarding writing and speaking as the websites or applications alone are not enough to provide necessary assessment and practice as 'human touch' and 'interaction with more proficient learners' is also needed. In addition to this, the reliability of the service provided by the applications and websites was also stressed by the participants as one of the disadvantages. It is a common fact that applications and websites can 'come and go' quickly and/or stop providing free service after getting popular among learners. Moreover, it appears to be rather difficult to determine which websites or application can provide reliable service. As a solution to this issue, teachers and learners can work together to inform each other about the Internet resources that provide similar services so that if any of these fail to work, they can continue using the alternative ones. In addition, it is due to note that teachers and parents should provide support and guidance to learners so that they can benefit from technology and make more connections between learning in and outside the classroom (Lewis, 2017; Lewin & Charania, 2018). The need for this support has been voiced by learners clearly stating that they need teachers as they need to be informed on useful technological resources and how to use these resources (Godwin-Jones, 2011; Lai, Yeung, & Hu, 2016).

FUTURE RESEARCH DIRECTIONS

It is due to note that the current study is limited in various aspects. It must be acknowledged that data were collected through participants' responses to the online questionnaire and semi-structured interview questions. It was believed that the participants provided true responses to the questions based on their own experiences. However, the responses of the participants might be different from what they really do or practice in real life. Therefore, it must be stated that since the findings of the study are only based on self-report of the participants who responded to the online survey, these findings might not be generalizable to all contexts. However, the findings can be transferable to other similar situations. Although during the interviews the participants' responses to the online questionnaire and the interview questions were compared and contrasted to determine what they matched, it was not possible for the questionnaire data. It is also due to note that the discussion of the findings of the study based on the online questionnaire and the responses obtained during interviews are impressionistic and based on the subjective responses. Therefore, further research can focus on which applications and websites learners actually use when they wish to assess their language skills. This can be done through tracking the participants' actions on their mobile devices and their computers. Moreover, they can also be asked to keep a journal on which they note down what they do in terms of assessing their knowledge and skills.

CONCLUSION

The current study included 500 university EFL learners in the intensive English programs in Turkey. The participants were directed to an online questionnaire in which they were asked to indicate the ICT applications and websites t employed informally outside the classroom and to explain how they used these applications and websites. The data were collected through *Google Forms* and analyzed using IBM SPSS Statistics 24. The questions required the participants to indicate which applications they used informally outside the classroom to assess their language skills and how they used these applications. Therefore, the findings were based on participants' perceptions and experience. The findings of the study indicated that the participants employed websites such as *ELLLO,* which provide freely available videos supported with online questions in order to improve and assess their listening skills and the websites with grammar questions and exercises to check their knowledge of the newly discussed items in the classrooms such as *The Grammar Aquarium* and *Road to Grammar*. The participants also created their own assessment tools especially for vocabulary through using flashcard creation applications and websites and benefited from applications on their smartphones. However, it was also determined that the participants had fewer options regarding speaking and writing assessment activities. Regarding the participants' choice of the applications that they could use to assess themselves, several reasons were provided us the cost of the application and easy access and use.

REFERENCES

Asri, A. N. (2019). Designing a 21st century assessment in EFL learning context. *KnE Social Sciences*, *3*(10), 335–348. doi:10.18502/kss.v3i10.3915

Brown, H. D., & Abeywickrama, P. (2010). *Language assessment: Principles and classroom practices* (2nd ed.). New York, NY: Pearson Education, Inc.

Brown, J. D. (1997). Computers in language testing: Present research and some future directions. *Language Learning & Technology*, *1*(1), 44–59. Retrieved from https://www.lltjournal.org/item/2234

Bunce, D. M., VandenPlas, J. R., & Havanki, K. L. (2006). Comparing the effectiveness on student achievement of a student response system versus online WebCT quizzes. *Journal of Chemical Education*, *83*(3), 488–493. doi:10.1021/ed083p488

Cambly. (2019). Retrieved from https://www.cambly.com

Cheng, L., & Fox, J. (2017). *Assessment in the language classroom: Teachers supporting student learning*. London: Palgrave. doi:10.1057/978-1-137-46484-2

Cheung, S. K. S. (2019). A study on the university students' use of open educational resources for learning purposes. In S. K. S. Cheung, J. Jiao, L.-K. Lee, X. Zhang, K. C. Li, & Z. Zhan (Eds.), *Technology in education: Pedagogical innovations. ICTE 2019*. Singapore: Springer. doi:10.1007/978-981-13-9895-7_13

Chinnery, G. M. (2006). Emerging technologies: Going to the MALL: Mobile assisted language learning. *Language Learning & Technology*, *10*(1), 9–16. Retrieved from https://scholarspace.manoa.hawaii.edu/bitstream/10125/44040/1/10_01_emerging.pdf

Combe, C., Folse, K., & Hubley, N. (2007). *A practical guide to assessing English language learners*. University of Michigan Press.

Creswell, J. W. (2005). *Educational research: Planning, conducting, and evaluating quantitative and qualitative research* (2nd ed.). Merrill Prentice Hall.

Cunningham, G. K. (1997). *Assessment in the classroom: Constructing and interpreting texts*. Washington, DC: The Falmer Press.

Daniela, L. (Ed.). (2019). Didactics of smart pedagogy: Smart pedagogy for technology enhanced learning. Cham: Springer.

DeSouza, E., & Fleming, M. (2003). A comparison of in-class and online quizzes on student exam performance. *Journal of Computing in Higher Education*, *14*(2), 121–134. doi:10.1007/BF02940941

Djoub, Z. (2017). Assessment literacy: Beyond teacher practice. In R. Al-Mahrooqi, C. Coombe, F. Al-Maamari, & V. Thakur (Eds.), *Revisiting EFL assessment: Critical perspectives* (pp. 9–27). Cham: Springer. doi:10.1007/978-3-319-32601-6_2

Dumova, T. (2012). The usability of online quizzes: Evaluating student perceptions. In S. Kelsey & K. St. Amant (Eds.), *Computer-mediated communication: Issues and approaches in education* (pp. 50–61). Hershey, PA: IGI Global. doi:10.4018/978-1-61350-077-4.ch004

Falsgraf, C. (2009). The ecology of assessment. *Language Teaching*, *42*(4), 491–503. doi:10.1017/S0261444808005570

Gareis, C. R., & Grant, L. W. (2015). *Teacher-made assessments: How to connect curriculum, instruction, and student learning* (2nd ed.). New York, NY: Routledge. doi:10.4324/9781315773414

Ghouali, K., & Benmoussat, S. (2019). Investigating the effect of social media on EFL students' written production: Case of third-year EFL students at Tlemcenz University, Algeria. In *Arab World English Journal, May 2019 Chlef University International Conference Proceedings* (pp. 24-39). doi:10.24093/awej/Chief1.3

Godwin-Jones, R. (2011). Emerging technologies: Autonomous language learning. *Language Learning & Technology*, *15*(3), 4–11. Retrieved from http://llt.msu.edu/issues/october2011/emerging.pdf

Golonka, E. W., Rowles, A. R., Frank, V. M., Richardon, D. L., & Freynik, S. (2014). Technologies for foreign language learning: A review of technology types and their effectiveness. *Computer Assisted Language Learning*, *27*(1), 70–105. doi:10.1080/09588221.2012.700315

Hung, Y. (2019). Bridging assessment and achievement: Repeated practice of self-assessment in college English classes in Taiwan. *Assessment & Evaluation in Higher Education*, *44*(8), 1191–1208. doi:10.1080/02602938.2019.1584783

Johnson, B. C., & Kiviniemi, M. T. (2009). The effect of online chapter quizzes on exam performance in an undergraduate social psychology. *Teaching of Psychology*, *36*(1), 33–37. doi:10.1080/00986280802528972 PMID:20046908

Katz, A., & Gottlieb, M. (2013). Assessment in the classroom. In C. A. Chapelle (Ed.), *The Encyclopedia of Applied Linguistics*. Oxford, UK: Wiley-Blackwell. doi:10.1002/9781405198431.wbeal0042

Kibble, J. (2007). Use of unsupervised online quizzes as formative assessment in a medical physiology course: Effects of incentives on student participation and performance. *Advances in Physiology Education*, *31*(3), 253–260. doi:10.1152/advan.00027.2007 PMID:17848591

Lai, C. (2015). Perceiving and traversing in-class and out-of-class learning: Accounts from foreign language learners in Hong Kong. *Innovation in Language Learning and Teaching*, *9*(3), 265–284. doi:10.1080/17501229.2014.918982

Lai, C., & Gu, M. (2011). Self-regulated out-of-class language learning with technology. *Computer Assisted Language Learning*, *24*(4), 317–335. doi:10.1080/09588221.2011.568417

Lai, C., Hu, X., & Lyu, B. (2018). Understanding the nature of learners' out-of-class language learning experience with technology. *Computer Assisted Language Learning*, *31*(1–2), 114–143. doi:10.1080/09588221.2017.1391293

Lai, C., Yeung, Y., & Hu, J. (2016). University student and teacher perceptions of teacher roles in promoting autonomous language learning with technology outside the classroom. *Computer Assisted Language Learning, 29*(4), 703–723. doi:10.1080/09588221.2015.1016441

Lewin, C., & Charania, A. (2018). Bridging formal and informal learning through technology in the twenty-first century: Issues and challenges. In J. Voogt, G. Knezek, R. Christensen, & K.-W. Lai (Eds.), *Second Handbook of Information Technology in Primary and Secondary Education* (pp. 199–215). Cham: Springer. doi:10.1007/978-3-319-71054-9_13

Lewis, G. (2017). *Learning technology*. Oxford, UK: Oxford University Press.

Literacynet. (2019). Retrieved from http://literacynet.org/cnnsf/archives.htm

Matsuno, S. (2009). Self-, peer-, and teacher-assessments in Japanese university EFL writing classrooms. *Language Testing, 26*(1), 75–100. doi:10.1177/0265532208097337

McMillan, J. H. (2010). The practical implications of educational aims and contexts for formative assessment. In H. E. Andrade & G. J. Cizek (Eds.), *Handbook of formative assessment* (pp. 41–58). New York, NY: Routledge.

Miller, M. D., Linn, R. L., & Gronlund, N. E. (2009). *Measurement and assessment in teaching* (10th ed.). Pearson Education Inc.

Naimi, Y. (2018). Teachers' conceptions of assessment in an ESP context. In S. Hidri (Ed.), *Revisiting the assessment of second language abilities: From theory to practice* (pp. 175–194). Cham: Springer. doi:10.1007/978-3-319-62884-4_9

Pérez-Solà, C., Herrera-Joancomartí, J., & Rifà-Pous, H. (2018). On improving automated self-assessment with Moodle quizzes: Experiences from a Cryptography course. In E. Ras & A. E. Guerrero Roldán (Eds.), *TEA 2017, CCIS 829* (pp. 176–189). Cham, Switzerland: Springer. doi:10.1007/978-3-319-97807-9_14

Sockett, G. (2014). *The online informal learning of English*. New York, NY: Palgrave Macmillan. doi:10.1057/9781137414885

Sundqvist, P., & Kerstin Sylvén, L. (2016). *Extramural English in teaching and learning: From theory and research to practice*. New York, NY: Palgrave Macmillan. doi:10.1057/978-1-137-46048-6

Toffoli, D., & Sockett, G. (2015). University teachers' perceptions of online informal learning of English (OILE). *Computer Assisted Language Learning, 28*(1), 7–21. doi:10.1080/09588221.2013.776970

Wang, S., & Smith, S. (2013). Reading and grammar learning through mobile phones. *Language Learning & Technology*, *17*(3), 117–134. Retrieved from http://llt.msu.edu/issues/october2013/wangsmith.pdf

Woit, D., & Mason, D. (2000). Enhancing student learning through on-line quizzes. In S. Haller (Ed.), *Proceedings of the thirty-first SIGCSE technical symposium on Computer science education* (pp. 367–371). New York, NY: ACM. 10.1145/330908.331887

Wright, W. E. (2015). *Foundations for teaching English language learners: Research, theory, policy, and practice*. Philadelphia: Caslon Publishing.

ADDITIONAL READING

Ferlazzo, L., & Sypnieski, K. H. (2018). *The ELL teacher's toolbox: Hundreds of practical ideas to support your students*. San Francisco, CA: Jossey-Bass. doi:10.1002/9781119428701

Navarre, A. (2019). *Technology-enhanced teaching and learning of Chinese as a foreign language*. New York: Routledge.

Stanley, G. (2013). *Language earning with technology: Ideas for integrating technology in the classroom*. Cambridge: Cambridge University Press.

Walker, A., & White, G. (2013). *Technology enhanced language learning: Connecting theory and practice*. Oxford: Oxford University Press.

KEY TERMS AND DEFINITIONS

Applications and Websites: Mobile and computer software and websites that can be used to assess English skills.

Informal Activities: Learners' activities online to learn and assess their English without the guidance of their teachers.

Online Self-Assessment: Learners' assessment of their skills using applications and websites on electronic devices generally through online quizzes and exercises.

APPENDIX

INTERVIEW QUESTIONS

1. What do you think are the major benefits/advantages of using applications and websites to assess your English?
2. What do you think are the major drawbacks/disadvantages of using applications and websites to assess your English?
3. How do you use applications and websites to assess your English grammar?
4. How do you use applications and websites to assess your English vocabulary?
5. How do you use applications and websites to assess your listening skill in English?
6. How do you use applications and websites to assess your reading skill in English?
7. How do you use applications and websites to assess your writing skill in English?
8. How do you use applications and websites to assess your speaking skill in English?
9. Would you like to add anything or raise any issue regarding the use of websites and applications for self-evaluation?

Chapter 6
An Investigation Into EFL Pre-Service Teachers' Out-of-Class Self-Regulated Learning Experiences With ICT

Saadet Korucu-Kis
https://orcid.org/0000-0002-1128-1747
Necmettin Erbakan University, Turkey

ABSTRACT

Self-regulated learning is one of the essential skills learners need to possess in times of rapid technological advancements to comply with the requirements of the modern world. Indeed, today's youth are assumed to be able to set their own goals, seek opportunities to attain these goals, and manage their learning environments to control their own learning owing to the omnipresence of information and communication technology (ICT) tools in every area of their lives including and especially in their educational lives. In this respect, this study aims to find out whether first-year English as a Foreign Language (EFL) pre-service teachers really make use of ICT tools to regulate their formal learning experiences in out-of-class settings. Results reveal that pre-service teachers most often fail in bridging in-class and out-of-class learning, and they generally engage in low-level self-regulated learning experiences. Implications arising from these findings are highlighted, and some future research possibilities are discussed.

DOI: 10.4018/978-1-7998-2116-8.ch006

Copyright © 2020, IGI Global. Copying or distributing in print or electronic forms without written permission of IGI Global is prohibited.

INTRODUCTION

With the advent of ICT, the scope of foreign language learning has extended beyond the limitations of traditional learning communities and classrooms. The emergence of such sophisticated tools as laptops, tablets and smartphones has provided learners with not only novel ways to learn the target language, but also flexibility of learning anywhere and anytime. Indeed, many successful foreign language learners ascribe their good performance in the target language to their endeavors outside of formal learning settings (Lamb, 2002). This being the case, the notion of out-of-class self-regulated learning with ICT has started to be in the spotlight of both researchers and practitioners in recent years.

Defined as "an active, constructive process whereby learners set goals for their learning and then attempt to monitor, regulate, and control their cognition, motivation, and behavior, guided and constrained by their goals and the contextual features of the environment" (Pintrich, 2000, p. 453), self-regulated learning has been an important concept in educational contexts. Helping learners in taking control of their own learning (Zimmerman, 1995), it has been promoted as an essential quality that should be developed in language learners (Lai & Gu, 2011; Wang, Hu, Zhang, Chang & Xu, 2012).

Out of class learning is one of the significant predictors of self-regulated learning and it has a complementary role for in-class learning (Lai & Gu, 2011). Benson (2011) suggests that self-regulated learning activities outside the classroom help learners manage their learning better as they become more independent in their decision-making with respect to their learning experiences. Consequently, several researchers have searched for new ways to increase self-regulated learning practices among English as a foreign language (EFL) learners. Providing learners with a variety of activities and authentic materials, increasing motivation and interaction, and supporting individualized instruction (Chapelle, 2000; Lee, 2000), ICT is thus viewed as one of the promising means of encouraging self-regulated language learning outside the classrooms. Accordingly, researchers (Lai, 2013, 2015; Lai, Zhu & Gong, 2015) suggest that supporting learners to self-direct their language learning beyond formal contexts is very important to exploit the potential of technology for language learning gains.

This issue is of particular relevance to countries like Turkey, where English is learnt and taught as a foreign language and learners' exposure to the target language is limited to classroom teaching. Similarly, existing literature (e.g. Nakata, 2019; Richards, 2015) also suggests that it becomes almost a necessity in such contexts to engage learners in different out-of-class activities to maximize their proficiency in English. (Nakata, 2019; Richards, 2015). Although numerous studies can be found regarding EFL learners' engagement in out-of-class language learning experiences with ICT, little is known about EFL pre-service teachers' ICT-integrated out-of-class study to regulate their in-class language learning. In an attempt to fill in this gap, the present study investigates whether first-year EFL pre-service teachers experience out-of-class self-regulated learning with modern ICTs to direct their formal language learning experiences. Given the fact that they are going to be teachers of English language in a foreign language context (Turkey) wherein they need to have an intrinsic drive to continue language learning on their own, the author believes that it becomes crucial to investigate whether pre-service teachers use ICTs to self-regulate their learning. Against this backdrop, the present study was guided by the following research questions:

1. Do first-year EFL pre-service teachers engage in out-of-class language learning experiences with ICT? If yes, in what ways?
2. Which ICT tools do EFL pre-service teachers use to self-regulate their formal language learning experiences outside the classroom?
3. What kind of self-regulated learning strategies do EFL pre-service teachers use ICT tools for?

It is assumed that this study will be helpful for the would-be teachers because understanding the nature of EFL pre-service teachers' use of ICT to self-monitor their out-of-class language learning is the key step toward encouraging and promoting their self-regulated uses of ICT for better improving their proficiency in English (Lai, 2013).

REVIEW OF LITERATURE

Theoretical Framework

The concept of self-regulation has been derived from Bandura's social cognitive theory, which posits that learning takes place through the constant interplay between personal, social and environmental variables. According to Bandura (1986), self-regulation refers to controlling one's behavior to reach desired objectives and it consists of three interconnected stages: self-observation, self-judgment, and self-reaction. Through these stages, learners observe and evaluate their behaviors, and respond to the results of the outcomes. If learners attain their goals, they feel competent and motivated to work towards attaining their future objectives. However, if they have doubts over their efficacy in achieving intended outcomes, their motivation may decline.

Since its introduction, the construct of self-regulation has continued to expand to other fields. Dinsmore, Alexander and Loughlin (2008) state that Zimmerman, Schunk and colleagues were among the first scholars to apply the concept in academic settings. According to Zimmerman (2002), "self-regulation refers to self-generated thoughts, feelings, and behaviors that are oriented to attaining goals" (p.65). It is also suggested as a process in which learners are "metacognitively, motivationally, and behaviorally active participants in their own learning process" (Zimmerman, 1989, p.329). In line with this, Winne (1995) indicates that self-regulated learners engage in educational activities with a conscious awareness of their beliefs, knowledge and behaviors. They set goals, use strategies, sustain their motivation drawing on their achievements, regulate their behaviors when they get unexpected outcomes and take great responsibility for reaching their goals. In this respect, Zimmerman (2000) notes that self-regulation is not a mental aptitude; it is the transformation of mental abilities into academic skills through self-directed endeavors.

In the existing literature, a number of self-regulated learning models can be found. According to Pintrich (2000), there are four assumptions common to all them: (a) learners are active constructors of knowledge, (b) learners can regulate their thoughts, actions and motivation, (c) learners can control their thoughts, actions and motivation to reach their aims and (d) self-regulatory activities have an intermediary role between personal and environmental factors and academic success or performance. However, among these, Zimmerman's Social Cognitive Model of Self-Regulation (1998) and Pintrich's General Framework for Self-Regulated Learning Model (2000) are the most commonly used models related to self-regulated learning.

Building on Bandura's (1986) cyclical model of self-regulation, Zimmerman (2000) suggests that the cycle of self-regulated learning consists of three phases: "forethought, performance or volitional control, and self-reflection processes" (p. 16). In the forethought stage, learners set their learning goals, identify the activities to be undertaken, and plan the order and timing of the selected tasks to attain them. The performance stage involves both performing the activities specified in the previous phase and monitoring learning process to optimize one's strengths and weaknesses to attain learning goals. In the self-reflection stage, the overall learning process is evaluated. Learners compare their performances with pre-established standards or engage in an attribution process to determine the causes of their achievements and failures. These reflections affect learners' reactions towards their learning experiences and in turn, influence the future forethought processes. In his model, Pintrich (2000) also suggests the same phases as Zimmerman (2000), but breaks down the second stage into cognitive monitoring, and control and regulation phases. He states that the control phase allows learners to choose appropriate cognitive strategies which can positively influence learners' learning processes and performances.

As the models imply, self-regulated learning also involves using several strategies to accomplish learning goals. Zimmerman and Martinez-Pons (1990) define self-regulated learning strategies as "actions directed at acquiring information or skill that involve agency, purpose (goals), and instrumentality self-perceptions by a learner" (p.615). They also list 14 strategies that self-regulated learners make use of: "organizing and transforming…, rehearsing and memorizing…, goal setting and planning…, self-evaluating and self-consequating…, seeking information…, record keeping and monitoring…, environmental structuring…, seeking social assistance…, and reviewing academic materials" (p. 51). Based on his model, Pintrich (1999) puts self-regulated learning strategies into three main categories as cognitive, metacognitive and resource management strategies. Cognitive strategies encompass rehearsing, elaborating, and organizing and they are used to learn, understand and remember the content. Metacognitive strategies refer to awareness about cognitive processes and they are employed to plan, monitor and modify cognition (Pintrich & De Groot, 1990). Resource management strategies concern the appropriate use of sources in the learning environment and include such strategies as managing time, organizing the study environment, and seeking help from more knowledgable others.

Because the primary aim of the present study is to investigate EFL pre-service teachers' out-of-class self-regulated learning experiences and specifically to find out which strategies they use ICTs for, the study draws both on Zimmerman's (1990) and Pintrich's (1999) theories. Given that undergraduate students of the 21st century are required to be lifelong and self-regulated learners, and ICT tools can serve this purpose allowing learners to go beyond the geographical limitations of their classrooms and independently regulate their cognition, actions and intrinsic incentives, both theories can provide insights into which strategies EFL pre-service teachers utilize from and whether they actively take part in their learning processes, and manage and direct their learning experiences.

Self-Regulated Learning in EFL Contexts

Self-regulated learning has been of growing interest to researchers in the discipline of applied linguistics for almost two decades owing to its promising features such as increasing language proficiency (Andrade & Bunker, 2009; Farsani, Beikmohammadi & Mohebbi, 2014; Seker, 2016), decreasing language anxiety (Martirossian & Hartoonian 2015) and developing autonomy (Hauck, 2005; Leaver, Ehrman & Shekhtman, 2005). Since it is very well established that acquiring a new language requires continuous exposure to target language input and ample opportunities for outputting the language, it has become more of an interest in foreign language contexts, where learners have limited opportunities to communicate in the target language. Research (e.g. Kennedy & Levy, 2009) hence suggests that learners in such environments need to be self-directed and manage their learning environments in a way that maximizes the opportunities for authentic language use.

Although self-regulation is a skill that is possessed by everyone, its level of mastery varies among individuals (Zimmerman, 2000). It can be taught (Bartolome & Steffens, 2011; Rozendaal, Minnaert, & Boekaerts, 2003) and developed in learners through constant practice and feedback (Seker & Dincer, 2016). Moreover, compared to poor self-regulators, since good self-regulators "set better learning goals, implement more effective learning strategies, monitor and assess their goal progress better, establish a more productive environment for learning, seek assistance more often when it is needed, expend effort and persist better, adjust strategies better, and set more effective new goals when present ones are completed" (Zimmerman & Schunk, 2008, p.1), it becomes highly important to have such kind of language learners in countries like Turkey to obtain better language gains. According to Orhon (2018), although Turkish students start learning English beginning from the lower levels of primary school and continue till the end of their high-school education, they cannot use it successfully despite many initatives (i.e. redesigning curricula, training inservice teachers, providing course materials) undertaken by the Ministry of National Eduation. Consequently, like many other researchers (Benson, 2011; Lai, 2011) around the world, she points to the significance of different learning environments in which language learners actively take part in their own learning.

Richards (2015) points out that although learning occurs both inside and outside the classroom, much attention has been paid to classroom-based teaching until not long ago. According to Benson (2011), the main reason for this disparity results from the fact that classrooms are viewed as the natural places for learning to occur. That said, Barbee (2013) states that such an assumption is problematic because learning continues outside of formal education institutions, as well. In a similar line of research, Richards (2009) indicates that "it has taken us a while to realize that while good teaching is no less important than ever, today's learners are not as dependent on classroom-based learning and teaching as they used to be" (p. 10). This suggests that learners resort to alternative resources to regulate their language learning experiences in informal settings and technology-supported learning environments provide venues for learners to practice the target language (McLoughlin & Lee, 2010).

ICT for Self-Regulated Language Learning in Out-Of-Class Settings

The paradigm shift from behaviorism to sociocultural view in learning has changed the role of students from passive recipients of content to life-long learners who actively construct their own knowledge. Learning through memorization and repetition has given way to learning through discovering, thinking, inquiring, creating, collaborating and problem-solving. According to Majumdar (2015) shifting the emphasis from teacher-centered to more student-centered activities, the new paradigm has empowered students to take more responsibility to regulate and control their learning. Enabling students to reach a world beyond the boundaries of their classrooms to access information in multiple ways, ICT has thus easily penetrated into students' out-of-class study.

Indeed, ICT has an important role in allowing out-of-class learning experiences to enjoy great popularity recently. An increasing number of studies (e.g. Belanger, 2011; Straka, 2004) suggest that most learning takes place outside of formal learning environments. Likening learning to an iceberg, Rogers (2014) states that while the tip of the iceberg stands for the formal learning, the part under the water represents the informal learning. He further states that the unseen part is more extensive and influential than what is seen. With a wide array of digital technologies like computers, communication services, the Internet, web services and numerous applications, ICT-mediated out-of-class learning hence assumes an important place in self-regulated language learning activities since it has the potential to help learners regulate their cognitive, meta-cognitive and behavioral processes as well as their learning environments (Kitsantas, 2013; Lai, 2013).

ICT provides language learners with venues and resources through which they can (a) combine learning and entertainment (Çelik, Arkın & Sabriler, 2012), (b) have access to authentic materials, (c) interact with supportive learning communities and authentic audience (Egbert, Chao, & Hanson-Smith, 1999), (d) address to their learning needs in accordance with their styles, proficiency and pace, and (e) receive high-quality feedback (Zhao & Lai, 2007). Hence, it is not incidental that many successful language learners attribute their achievement in the target language to their technology-supported self-initiated language learning experiences beyond the classroom (Bernacki, Aguilar & Byrnes, 2011).

So far, a number of studies have been carried out to investigate the impact of ICT on self-regulated language learning skills. For example, in an attempt to find out to what extent 279 Hong Kong students' out-of-class ICT use influenced their self-regulated language learning skills, Lai and Gu (2011) found that half of the participants were positive about their ICT-integrated learning experiences in helping them reach their language learning goals and sustain their motivation. In like manner, Richards (2009) stated that compared to their counterparts in countries like Italy or Japan, language learners in northern European countries performed better in the target language as a result of authentic exposure to English language through undubbed programs. In the Turkish context, Çelik, Arkın and Sabriler (2012) examined 399 EFL learners' self-initiated uses of ICT for language learning purposes through a survey study. The researchers found that learners mostly engaged in self-regulated learning activities with ICT to improve their listening, vocabulary and writing skills. The learners also reported that ICT helped them sustain their motivation and allowed them to interact with others for practicing purposes. Similarly, inquiring about 777 university students' self-regulated language learning experiences with ICT, Şahin-Kızıl and Savran (2016) found that participants felt positive about the fact that ICT provided them with individualized content, anxiety-free learning environments and a myriad of authentic materials.

In view of all these observations, it can be said that there is a close relationship between ICT and self-regulated language learning. Accordingly, Lai and Gu (2011) state that while ICT-supported learning environments provide learners with best opportunities to manage their learning experiences in foreign language contexts, the potential of ICT can best be unlocked in self-regulated learning environments. It is thus essential to encourage and support language learners to regulate their learning experiences with ICT in order to fulfill this potential. To do so, the first step lies in understanding whether and in what ways learners use ICT tools to engage in self-regulated language learning activities outside the classroom and this explains the motive behind carrying out this study.

METHODOLOGY

A cross-sectional survey design was adopted in the present study. This method enables researchers to investigate current opinions of a population at one point in time (Creswell, 2012). The data obtained describes the nature of existing conditions and it may subsequently lead researchers to either a prospective or retrospective inquiry (Cohen, Manion & Morrison, 2000).

Context and Participants

This study was carried out in the spring term of the 2018-2019 academic year in an English Language Teacher Education (ELTE) program of a state university in Turkey. The curricula of teacher education programs in Turkey are designed by the Council of Higher Education (CoHE) and comprise both campus-based courses and field-based experiences. In the first year of ELTE programs, pre-service teachers are required to take language skills courses such as Reading Skills I-II, Writing Skills I-II, Oral Communication Skills I-II, Listening and Pronunciation Course I-II and the Structure of English Language Course. The following years focus more on methodological content and pre-service teachers start practicum in the final year.

As to the status of technology use in Turkey, the first initiatives with respect to the integration of technology into educational contexts date back to 1980s. Several small- and large-scale projects have been carried out since then to address the demands of modernizing and technologizing educational settings. To illustrate, the Turkish Ministry of National Education initiated one of the world's largest-scale educational technology projects in 2010. Valued at US$1.8 billion, the project of Movement of Enhancing Opportunities and Improving Technology (FATIH) aimed to enhance the quality of educational practices by establishing hardware and software infrastructure, providing educational e-content, training in-service teachers and ensuring the effective use of ICT tools in schools (Pouzevara, Dincer, Kipp & Sariisik, 2014). In addition, according to the reports of the Turkish Statistical Institute (2018), the rate of computer and internet use among people aged 16-74 years was 59.6% and 72.9% respectively. The main purposes of using ICT tools among those who use the Internet include participating in social networks (84.1%), using video-sharing sites (78.1%) and telephoning over the Internet (69.5%). The availability rate of desktop/portable computers and cellphones/smartphones in households was 85.5% and 98.7% respectively. Compared to previous years, these data show that ICT use is rapidly increasing in Turkey.

The participants of the study were a convenience sample of 78 first-year pre-service teachers. 67.9% of the participants were females and 32.1% of them were males. Their ages ranged between 18 and 23 with the exception of 3 students at the ages of 24-35 years.

Data Collection and Procedure

Data for this study came from two sources: (a) the survey of self-regulated learning with technology at the University (Yot-Dominguez & Marcelo, 2017) validated for the Turkish context by the researcher and (b) 4 background questions inquiring about participants' gender, age and engagement in language practicing activities with ICT tools outside the classroom.

Permission to use the survey was granted via email. Prior to administering the survey, it was tested for validity and reliability with 180 pre-service EFL teachers out of the possible population of 78 first-year pre-service teachers. To these ends, the Statistical Package for Social Sciences software (SPSS) 25.0 was used. First, the Kaiser-Meyer-Olkin Measure of Sampling Adequacy (KMO) and Bartlett's Test of Sphericity were calculated to determine the suitability of the data for factor analysis. The KMO value was .793 which indicated an acceptable degree of common variance. The value of the Bartlett's Test for Sphericity was found to be significant (x^2=5105.140, $p<.001$). Then, a principal component exploratory factor analysis with varimax rotation was performed to determine the construct validity of the survey. Factors with eigenvalues over 1.0 were identified and nine factors were obtained. The analysis results showed that all nine factors together accounted for 82.306% of the total variance. In order to measure the internal consistency of the factors, Cronbach's alpha coefficient was calculated. Each factor showed high internal consistency ($\alpha = 0{,}961; 0{,}920; 0{,}863; 0{,}834; 0{,}886; 0{,}888; 0{,}942; 0{,}905$ and $0{,}885$ respectively). The Cronbach's α value for the total scale was 0.870. Since the factors had loadings over .45 and had no overlaps, all survey items were retained. Last, LISREL 8.7 was used to perform a confirmatory factor analysis to validate the factor structure derived from the analysis. The results of the confirmatory factor analysis displayed that nine-dimensional model was well fit (x^2=718.78; df= 459; x^2/fd=1.56; RMSEA=0.056; GFI= 0.95; AGFI= 0.92; RMR= 0.077; SRMR= 0.052; CFI= 0.96; NFI= 0.95 and NNFI= 0.98). Since the survey was found valid and reliable, no change was made to the original survey. As a further note, the survey was not translated into Turkish, since the participants were considered to be moderately proficient users of English given that they were able to start studying in the ELTE program at issue after passing the university's English proficiency examination with a minimum score of 70 which equals to 72 in the TOEFL-IBT.

The survey (see Appendix) is a five-likert scale ranging from 1 (Never) to 5 (Always) and it can be used to find out what technologies are utilized by university students for self-regulated learning purposes and what self-regulated learning activities university students engage in to promote the learning process. It consists of 33 items built on Zimmerman's (1998) and Pintrich's (2004) self-regulated learning models. The items refer to different self-regulated learning strategies that may be employed using a variety of technologies. The survey comprises nine factors: share information, active presence, documentation and classification, superficial use with information processing, expansion and in-depth information, monitoring and feedback, personal management, self-evaluation and collaborative learning. Participation in the study was voluntary and the respondents (N=78) were informed about the purpose of the study before filling in the survey.

Data Analysis

The data obtained from the survey was analyzed by the software of SPSS. Since this study was carried out to display the current trends among EFL pre-service teachers with respect to their self-regulated learning experiences with technology, descriptive statistics (i.e. percentages, means and standard deviations) were used to analyze the data.

RESULTS

1. Do first-year EFL pre-service teachers engage in out-of-class learning experiences with ICT? If yes, in what ways?

This question garners general information about the participants' technology profiles. In order to respond to the first research question, the participants were requested to answer the following closed-ended background questions before filling in the Survey of Self-regulated Learning with Technology at the University: "Do you practice language skills outside the classroom?", "Do you use technology tools to practice language skills outside the classroom?", "How often do you practice language skills outside the classroom?" and "What ICT-supported activities do you engage in to improve your language skills?". Table 1 shows the percentages, means and standard deviations of responses.

As seen in Table 1, the majority of pre-service teachers report that they practice language skills outside the classroom and use technological tools in their out-of-class language learning experiences. Over half of them state that they often engage in language learning experiences outside the classroom and use ICTs to practice listening, vocabulary, speaking, writing and reading most. Grammar is the least studied of all beyond the classrooms.

2. Which ICT tools do first-year EFL pre-service teachers use to self-regulate their formal language learning experiences outside the classroom?

As indicated in Table 2, pre-service teachers use ICT tools in out-of-class settings to regulate their formal language learning experiences but their use is very restricted. When the listed technologies are examined, it can be seen that only the Internet (M=3.19, SD=.74), social networks (M=2.83, SD=1.06), communication tools (M=2.72, SD=.71) and repositories (M=2.69, SD=.47) have a moderate level of use (2.5<M<3.5). ICT tools like social markers and RSS, and management tools have almost no use among pre-service teachers.

3. What kind of self-regulated learning strategies do EFL pre-service teachers use ICT tools for?

As aforementioned, the curriculum of first-year English language teaching programs in Turkey is based on improving pre-service teachers' language proficiency. Therefore, they were asked to rate what kind of self-regulated learning activities they engage in in out-of-class settings to regulate their formal language learning.

Table 1. First-year EFL pre-service teachers' technology use profiles

		%	M	SD
Practicing language skills beyond the classrooms	Yes	94	1.94	.25
	No	6		
Using technology tools to practice language skills beyond the classrooms	Yes	92	1.92	.27
	No	8		
The frequency of practicing language skills outside the classroom	Never	-	3.59	.78
	Rarely	6		
	Sometimes	40		
	Usually	42		
	Always	12		
ICT-supported activities undertaken by the participants to improve their language skills				
	Online reading activities (e.g. articles, news, books, blogs)	50	1.5	.50
	Online writing activities (e.g. blogs, diaries, texting, emails)	59	1.59	.50
	Online speaking activities (e.g. conversation with fellow students, native speakers)	60	1.6	.49
	Online listening activities (e.g. watching videos, movies; listening to news, songs, podcasts)	77	1.77	.42
	Online vocabulary activities (e.g. using dictionaries, translation apps)	67	1.67	.47
	Online grammar activities	22	1.22	.42

Table 3 shows that pre-service teachers frequently use ICT tools that help them clarify concepts covered in class-hours (M=3.81, SD=.91). Namely, 65% of the pre-service teachers reported that "I use online translation tools or mobile translation apps to better understand articles written in other languages about course-related material" and 67% of the pre-service teachers stated that "I consult Wikipedia or online dictionaries when I need to clarify a matter or concept related to the material covered in class." This is followed by pre-service teachers' involvement in getting more in-depth information about the subjects covered in their programs (M=3.45, SD=.70). 88% of the pre-service teachers reported that "While studying, I review my instructors' presentations, which were created using PowerPoint, Impress, Prezi, etc."; 60% of the pre-service teachers stated that "While studying, I watch videos on YouTube, Vimeo, etc."; 54% of the pre-service teachers indicated that "I follow YouTube channels where videos related to course content are posted and only 17%

Table 2. ICT tools EFL pre-service teachers use for formal language learning experiences

	M	SD
Social Networks: Twitter, Facebook	2.83	1.06
Multimedia Resources: podcast, video	1.86	.83
Production and Storage Tools in the cloud: wikis & blogs, Dropbox, Google +.	2.26	.72
Repositories: Slideshare, Instagram, Pinterest, Issuu, Calameo, Youtube, iTunes, iVoox.	2.69	.47
Internet: Wikipedia or online dictionaries; translators; Google academic, Dialnet.	3.19	.74
Management tools: Cmap, MindManager; RefWorks, Mendeley; Viper.	1.47	.65
Communication Tools: WhatsApp, Line, Skype, Google Talk.	2.72	.71
Other Technology: specific apps (Kalkulilo, Whiteboard Lite) & organisers (Google Calendar, EverNote).	1.69	.85
Assessment Tools: ExamTime, Google Application Forms	1.71	.78
Social Markers & RSS: Delicious, Sage.	1.30	.60

of the student teachers expressed that "I follow blogs written by experts who post about matters related to course content." Active presence on the web was the third frequently attended activity by the pre-service teachers to regulate their in-class learning outside of the classrooms (M=2.97, SD=.94). 23% of the pre-service teachers reported that "I comment on the information provided by my instructors during class using Twitter, Facebook, etc."; 56% of the pre-service teachers state that "I share images of in-class activities using Instagram, Flickr, Pinterest, Facebook, Twitter, or

Table 3. Self-regulated learning activities EFL pre-service teachers engage in through ICTs

	M	SD
Share information	1.50	.55
Active presence	2.97	.94
Documentation and classification	1.71	.79
Superficial use with limited information processing	3.81	.91
Expansion and in-depth information	3.45	.70
Personal management	1.7	.58
Monitoring and feedback	1.95	.80
Self-evaluation	1.71	.78
Collaborative learning	2.66	.58

other similar platforms."; 55% of the pre-service teachers indicate that "I read news, information, and updates about course-related content that are shared on social media (Facebook, Twitter, Google+, etc.)" and 23% of the pre-service teachers expressed that "I am part of groups within social networks (LinkedIn, Facebook, Google+, etc.) that discuss and exchange information about topics related to the material covered in the courses that I am taking." Taking part in collaborative learning experiences is the fourth frequently attended activity by the pre-service teachers (M=2.66, SD=.58). 20% of the pre-service teachers reported that "I share study materials and resources with my classmates using Dropbox, Facebook, Google+, etc."; 23% of the pre-service teachers stated that "I complete assignments, in collaboration with my classmates, using tools such as wikis, Google Drive, Dropbox, etc."; 66% of the pre-service teachers indicated that "I interact with my classmates outside the classroom setting using mobile apps (WhatsApp, Viber, Tango, Line, etc.) exchanging information, clarifying concepts, etc.", and finally, 17% of the pre-service teachers express that "I communicate with my classmates using videoconference tools (Skype, Google Talk, FaceTime, TeamViewer, etc.) to clarify concepts or discuss topics related to the courses I am taking." However, out-of-class learning experiences which require higher order information processing skills such as sharing information (M=1.50, SD=.55), personal management (M=1.7, SD=.58), and self-evaluation (M=1.71, SD=.78) were the least undertaken activities by the pre-service teachersto regulate their learning.

DISCUSSION

Although learning takes place across various contexts, much research focused on learning activities undertaken in formal instructional settings until a few decades ago (Benson, 2011; Richards, 2015). However, particularly with the proliferation of ICT, increasing calls have been made for the promotion of out-of-class learning experiences which embrace numerous opportunities for learning. It has been suggested that successful language learners are those who actively make use of various settings for their learning (Borrero & Yeh, 2010; Hyland, 2004). It has also been pointed out that integration of in-class learning with out-of-class learning experiences results in higher academic achievements (Chan, 2016). In an attempt to contribute to this line of research, the present study was carried out.

Overall, the study found that first-year EFL pre-service teachers engaged in a number of self-iniated learning experiences to increase their proficiency in language skills using a number of ICT tools. In alignment with previous research (e.g. Çelik, Arkın & Sabriler, 2012; Eksi & Aydın, 2013; Manfred, 2012; Sundqvist & Sylven, 2014), the present study revealed that learners most frequently used ICTs to practice their listening skills and watching movies/videos, and listening to news/songs were the most commonly preferred activities undertaken by the pre-service teachers in out-of-class settings.

However, as to bridging their in-class learning–designed primarily to improve their proficiency in language skills and areas (i.e. reading, writing, listening, speaking, vocabulary and grammar)– with their out-of-class study, the findings showed that most of the pre-service teachers failed in making use of different technologies to regulate their learning experiences. Although engagement in out-of-class learning experiences can be highly beneficial for pre-service teachers to revise what is covered during class hours, the results indicated that pre-service teachers mostly preferred practicing language skills outside the classrooms through aforementioned entertainment activities rather than more serious learning experiences like creating e-books, taking part in webinars or using wikis to complete assigments. A possible explanation for these results may be pre-service teachers' lack of adequate knowledge about the educational potential of these different types of technologies as suggested by many in the literature (e.g. Korucu-Kis & Ozmen, 2019a; Korucu-Kis & Ozmen, 2019b; Lai & Gu, 2011; Winke & Goertler, 2008).

Another major finding of the study was related to pre-service teachers' ICT use for self-regulated learning purposes. The study indicated that pre-service teachers favored studying in collaboration with their peers and would like to have a presence on the network to share and exchange their ideas. These findings support those of Yot-Dominguez and Marcelo (2017) who stated that learners needed to have an active presence beyond the physical academic spaces to project their learning and to be able to learn collaboratively with their classmates since they preferred using digital technologies for a more connected learning model.

Lastly, in line with those of previous studies (Ertmer, 2005; Ibieta, Hinostroza, Labbe & Claro, 2017; Korucu-Kis & Ozmen, 2019a; Wozney, Venkatesh & Abrami, 2006; Yot-Dominguez & Marcelo, 2017), the study pointed out that pre-service teachers generally used ICTs for accomplishing low-level tasks such as searching, reading and translating, and only a small number of participants involved in higher level tasks such as monitoring and self-assessment to regulate their out-of-class learning activities. These findings with the former ones show that the strategies pre-service teachers use to regulate their learning experiences primarily address to the cognitive and resource management aspects of Zimmerman's (1990) and Pintrich's (1999) theories. The findings further support the ideas of Mirriahi and Alonzo (2015) who posit that despite the increasing variety of technologies, students' use of technology for learning has been unchanged for years.

CONCLUSION

Moving beyond simply finding out whether and for what purposes learners engage in out-of-class learning and what ICTs are employed for learning beyond the classrooms, the chapter also aimed to explore the ICTs and strategies learners utilize from to regulate their in-class learning. The findings provide important insights into the nature of ICT-supported self-regulated learning activities among pre-service teachers in the Turkish context. It has been revealed that pre-service teachers use ICTs mainly for fun activities to keep their interest in and motivation for practicing the target language outside the classroom. The findings have also shown that they do not use different types of technologies beyond the classrooms to further the academic content covered in their formal learning programs. Overall, the findings suggest that first-year EFL pre-service teachers largely fail to engage in self-regulated learning activities and use ICTs to promote their learning experiences.

Although it would be misleading to generalize these findings to other populations in different settings within the country given the descriptive nature of the study and the limited number of participants taking part in the study, the chapter has important suggestions for practice and research. To start with, the findings reveal the need for making pre-service teachers aware of the concept of self-regulated learning and assisting them in making use of available ICT resources to support their self-initiated practices to improve their learning. It is, therefore, recommended that teacher educators should explicitly train pre-service teachers on the use of ICTs for self-regulated learning purposes, and design learning environments in which pre-service teachers understand how and when they can utilize from ICT-integrated self-regulated learning experiences. Teacher educators should also bridge in-class and out-of-class learning because each of these learning contexts has complementary roles in promoting learner success. However, although teacher educators seem to have a key role in cultivating the idea of self-regulated learning in pre-service teachers, their awareness about various ICT tools and self-regulated learning may come into question. Therefore, in-service training programs should be organized for helping them acquaint themselves with self-regulated learning strategies and available digital technologies to promote the learning process. Moreover, higher education curricula should consider and practice the idea that learning is not limited to classrooms and there are a variety of resources and activities that can be capitalized on beyond the classrooms.

Since such findings are critical to understand how to support pre-service teachers to undertake quality out-of-class language learning experiences and encourage them to partake in tasks which require higher-order thinking skills, this study suggests further research to be conducted with larger sample sizes and employing qualitative methods (e.g. *a hermeneutical phenomenological approach)* to bring more breadth and depth to the issue in question. Given this study is a cross-sectional one, longitudinal and experimental studies can be carried out to find out whether and how ICT-embedded strategy training can help learners develop self-regulated learning habits and improve their in-class learning.

REFERENCES

Andrade, M. S., & Bunker, E. L. (2009). A model for self-regulated distance language learning. *Distance Education*, *30*(1), 47–61. doi:10.1080/01587910902845956

Bandura, A. (1986). *Social foundations of thought and action: A social cognitive theory*. Englewood Cliffs, NJ: Prentice-Hall, Inc.

Barbee, M. (2013). Extracurricular L2 input in a Japanese EFL context: Exposure, attitudes, and motivation. *University of Hawaii Second Language Studies Paper, 32*, 1–58. Retrieved from http://www.hawaii.edu/sls/wp-content/uploads/2014/09/Barbee-Matthew-Paper.pdf

Bartolomé, A., & Steffens, K. (2011). Technologies for self-regulated learning. In R. Carneiro, P. Lefrere, K. Steffens, & J. Underwood (Eds.), Self-regulated learning in technology enhanced learning environments: A European perspective (pp. 21-31). Rotterdam: Sense Publishers. doi:10.1007/978-94-6091-654-0_2

Belanger, P. (2011). *Theories in adult learning and education*. Opladen: Barbara Budrich.

Benson, P. (2011). Language learning and teaching beyond the classroom: An introduction to the field. In P. Benson & H. Reinders (Eds.), *Beyond the language classroom* (pp. 7–16). Basingstoke, UK: Palgrave Macmillan. doi:10.1057/9780230306790_2

Bernacki, M. L., Aguilar, A. C., & Byrnes, J. P. (2011). Self-regulated learning and technology enhanced learning environments: an opportunity-propensity analysis. In G. Dettori & D. Persico (Eds.), *Fostering self-regulated learning through ICT* (pp. 1–26). New York: Information Science Reference. doi:10.4018/978-1-61692-901-5.ch001

Borrero, N., & Yeh, C. J. (2010). Ecological English language learning among ethnic minority youth. *Educational Researcher*, *39*(8), 571–581. doi:10.3102/0013189X10389810

Çelik, S., Arkın, E., & Sabriler, D. (2012). EFL learners' use of ICT for self-regulated learning. *The Journal of Language and Linguistic Studies*, *8*(2), 98–118. Retrieved from http://www.jlls.org/index.php/jlls/article/view/133

Chan, H. W. (2016). Popular culture, English out-of-class activities, and learner autonomy among highly proficient secondary students in Hong kong. *Universal Journal of Educational Research*, *4*(8), 1918–1923. doi:10.13189/ujer.2016.040823

Chapelle, C. (2010). Evaluating computer technology for language learning. *Teachers of English as a Second Language of Ontario*, *36*(2), 56–67. doi:10.1177/1046878191222006

Cohen, L., Manion, L., & Morrison, K. (2007). *Research methods in education*. London: Routledge. doi:10.4324/9780203029053

Creswell, J. W. (2012). *Educational research: Planning, conducting, and evaluating quantitative and qualitative research*. Boston: Pearson.

Dinsmore, D. L., Alexander, P. A., & Loughlin, S. M. (2008). Focusing the conceptual lens on metacognition, self-regulation, and self-regulated learning. *Educational Psychology Review*, *20*(4), 391–409. doi:10.100710648-008-9083-6

Egbert, J., Chao, C., & Hanson-Smith, E. (1999). Compsuter-enhanced language learning environments. An overview. In J. Egbert & E. Hanson-Smith (Eds.), *CALL Environments* (pp. 1–13). Alexandria, VA: TESOL Publications.

Ekşi, G., & Aydın, H. (2013). What are the students doing out there? An investigation of out-of-class language learning activities. *AIBU Journal of Social Sciences, 13*(2), 191-210. Retrieved from https://dergipark.org.tr/download/article-file/154557

Ertmer, P. A. (2005). Teacher pedagogical beliefs: The final frontier in our quest for technology integration? *Educational Technology Research and Development*, *53*(4), 25–39. doi:10.1007/BF02504683

Farsani, M. A., Beikmohammadi, M., & Mohebbi, A. (2014). Self-regulated learning, goal-oriented learning, and academic writing performance of undergraduate Iranian EFL learners. *TESL-EJ, 18*(2), 1-19. Retrieved from http://www.tesl-ej.org/wordpress/issues/volume18/ej70/ej70a4/

Hauck, M. (2005). Metacognitive knowledge, metacognitive strategies and CALL. In J. L. Egbert & G. M. Petrie (Eds.), *CALL research perspectives* (pp. 65–86). New York: Lawrence Erlbaum Associates.

Hyland, F. (2004). Learning autonomously: Contextualizing out-of-class English language learning. *Language Awareness*, *13*(3), 180–202. doi:10.1080/09658410408667094

Ibieta, A., Hinostroza, J. E., Labbe, C., & Claro, M. (2017). The role of the internet in teachers' professional practice: Activities and factors associated with teacher use of ICT inside and outside the classroom. *Technology, Pedagogy and Education*, *26*(4), 425–438. doi:10.1080/1475939X.2017.1296489

Kennedy, C., & Levy, M. (2009). Sustainability and CALL: Factors for success in a context of change. *Computer Assisted Language Learning*, *22*(5), 445–463. doi:10.1080/09588220903345218

Kitsantas, A., & Dabbagh, N. (2011). The role of web 2.0 technologies in self-regulated learning. *New Directions for Teaching and Learning*, *2011*(126), 99–106. doi:10.1002/tl.448

Korucu-Kis, S., & Ozmen, K. S. (2019a). Exherent and Inherent Value Beliefs about Technology: Missing Pieces in the Puzzle of Technology Integration? *International Journal of Educational Technology*, *6*(1), 1–11. Retrieved from https://educationaltechnology.net/ijet/index.php/ijet/article/view/92

Korucu-Kis, S., & Ozmen, K. S. (2019b). S-TECHNO: An instructional design model for designing instructional technology courses. *International Journal of Education and Development Using Information and Communication Technology*, *15*(1), 99–117. Retrieved from http://ijedict.dec.uwi.edu/viewissue.php?id=53

Lai, C. (2013). A framework of developing self-directed technology use for language learning. *Language Learning & Technology*, *17*(2), 100–122. Retrieved from https://scholarspace.manoa.hawaii.edu/bitstream/10125/44326/1/17_02_lai.pdf

Lai, C. (2015). Perceiving and traversing in-class and out-of-class learning: Accounts from foreign language learners in Hong Kong. *Innovation in Language Learning and Teaching*, *9*(3), 265–284. doi:10.1080/17501229.2014.918982

Lai, C., & Gu, M. Y. (2011). Self-regulated out-of-class language learning with technology. *Computer Assisted Language Learning*, *24*(4), 317–335. doi:10.1080/09588221.2011.568417

Lai, C., Zhu, W. M., & Gong, G. (2015). Understanding the quality of out-of-class English learning. *TESOL Quarterly*, *49*(2), 278–308. doi:10.1002/tesq.171

Lamb, M. (2002). Explaining successful language learning in difficult circumstances. *Prospect*, *17*(2), 35–52.

Leaver, B. L., Ehrman, M., & Shekhtman, B. (2005). *Achieving success in second language acquisition*. Cambridge, UK: Cambridge University Press. doi:10.1017/CBO9780511610431

Lee, K. (2000). English teachers' barriers to the use of computer–assisted language Learning. *The Internet TESL Journal*, *6*(12). Retrieved from http://iteslj.org/Articles/Lee–CALLbarriers.html

Majumdar, S. (2015). *Emerging Trends in ICT for education & Training*. Retrieved from http://www.unesc o.org/fileadmin/up/emergingtrendsinicteducationandtraining.pdf

Manfred, W. M. (2012). Beliefs and Out-of-class language learning of Chinese-speaking ESL learners in Hong Kong. *New Horizons in Education*, *60*(1), 35–52. Retrieved from https://files.eric.ed.gov/fulltext/EJ974077.pdf

Martirossian, A., & Hartoonian, A. (2015). Lowering foreign language anxiety through self-regulated learning strategy use. *English Language Teaching*, *8*(12), 209–222. doi:10.5539/elt.v8n12p209

McLoughlin, C., & Lee, M. J. W. (2010). Personalised and self regulated learning in the Web 2.0 era: International exemplars of innovative pedagogy using social software. *Australasian Journal of Educational Technology*, *26*(1), 28–43. doi:10.14742/ajet.1100

Mirriahi, N., & Alonzo, D. (2015). Shedding light on students' technology preferences: Implications for academic development. *Journal of University Teaching & Learning Practice*, *12*(1), 1–14. Retrieved from https://pdfs.semanticscholar.org/3197/d55590e546ffa1e6c934cd11af3b14d0f042.pdf?_ga=2.116175622.1832193434.1563368940-1412330347.1563368940

Nakata, Y. (2019). Encouraging student teachers to support self-regulated learning: A multiple case study on prospective language teachers. *International Journal of Educational Research*, *95*, 200–211. doi:10.1016/j.ijer.2019.01.007

Orhon, Y. (2018). An investigation of out-of-class language activities of tertiary-level EFL learners. *Education Reform Journal*. *3*(1), 1-14. Retrieved from https://dergipark.org.tr/download/article-file/516353

Oxford, R. L., Rubin, J., Chamot, A. U., Schramm, K., Lavine, R., Gunning, P., & Nel, C. (2014). The learning strategy prism: Perspectives of learning strategy experts. *System*, *43*, 30–49. doi:10.1016/j.system.2014.02.004

Pintrich, P., & De Groot, E. V. (1990). Motivational and self-regulated learning components of classroom academic performance. *Journal of Educational Psychology*, *82*(1), 33–40. doi:10.1037/0022-0663.82.1.33

Pintrich, P. R. (1999). The role of motivation in promoting and sustaining self-regulated learning. *International Journal of Educational Research*, *31*(6), 459–470. doi:10.1016/S0883-0355(99)00015-4

Pintrich, P. R. (2000). The role of goal orientation in self-regulated learning. In M. Boekaerts, P. R. Pintrich, & M. Zeidner (Eds.), *Handbook of self-regulation* (pp. 451–502). SanDiego, CA: Academic. doi:10.1016/B978-012109890-2/50043-3

Pintrich, P. R. (2004). A conceptual framework for assessing motivation and self-regulated learning in college students. *Educational Psychology Review*, *16*(4), 385–407. doi:10.100710648-004-0006-x

Pouezevara, S., Dincer, A., Kipp, S., & Sariisik, Y. (2013). *Turkey's FATIH project: A plan to conquer the digital divide or a technological leap of faith?* Research Triangle Park, NC: RTI International & Education Reform Initiative (ERI). Retrieved from http://en.egitimreformugirisimi.org/issue/turkeys-fatih-project/

Richards, J. C. (2009, April). *The changing face of TESOL.* Plenary Speech at TESOL Convention and Exhibit, Denver, CO. Retrieved from http://www.professorjackrichards.com/wp-content/uploads/changing-face-ofTESOL.pdf

Richards, J. C. (2015). The changing face of language learning: Learning beyond the classroom. *RELC Journal*, *46*(1), 5–22. doi:10.1177/0033688214561621

Rogers, A. (2014). The classroom and the everyday: The importance of informal learning for formal learning. *Investigar Em Educação*, *2*(1), 7–34.

Rozendaal, J. S., Minnaert, A., & Boekaerts, M. (2003). Motivation and self-regulated learning in secondary vocational education: Information-processing type and gender differences. *Learning and Individual Differences*, *13*(4), 273–289. doi:10.1016/S1041-6080(03)00016-5

Sahin-Kizil, A. S., & Savran, Z. (2016). Self-regulated learning in the digital age: An EFL perspective. *Novitas-ROYAL*, *10*(2), 147–158. Retrieved from http://www.novitasroyal.org/Vol_10_2/4.Sahin%20&%20Savran.pdf

Seker, M. (2016). The use of self-regulation strategies by foreign language learners and its role in language achievement. *Language Teaching Research*, *20*(5), 600–618. doi:10.1177/1362168815578550

Seker, M., & Dincer, A. (2016). Teacher transformation as a basis for the promotion of selfregulated language strategies. *Literacy Information and Computer Education Journal*, *7*(1), 2206–2210. doi:10.20533/licej.2040.2589.2016.0291

Straka, G. A. (2004). *Informal learning: concepts, antagonisms and questions.* Retrieved from https://www.pedocs.de/volltexte/2014/9162/pdf/Straka_2004_Informal_learning.pdf

Sundqvist, P., & Sylven, L. K. (2014). Language-related computer use: Focus on young L2 English learners in Sweden. *ReCALL*, *26*(1), 3–20. doi:10.1017/S0958344013000232

Turkish Statistical Institute. (2018). *Information and Communication Technology (ICT) usage survey on households and individuals.* Retrieved from http://www.turkstat.gov.tr/PreHaberBultenleri.do?id=21779

Wang, C., Hu, J., Zhang, G., Chang, Y., & Xu, Y. (2012). Chinese college students' self-regulated learning strategies and selfefficacy beliefs in learning English as a foreign language. *Journal of Research in Education*, *22*(2), 103–135. Retrieved from https://files.eric.ed.gov/fulltext/EJ1098422.pdf

Winke, P., & Goertler, S. (2008). Did we forget someone? Students' computer access and literacy for CALL. *CALICO Journal*, *25*(3), 482–509. doi:10.1558/cj.v25i3.482-509

Winne, P. H. (1995). Inherent details in self-regulated learning. *Educational Psychologist*, *30*(4), 173–187. doi:10.120715326985ep3004_2

Wozney, L., Venkatesh, V., & Abrami, P. C. (2006). Implementing computer technologies: Teachers' perceptions and practices. *Journal of Technology and Teacher Education*, *14*(1), 120–173.

Yot-Domínguez, C., & Marcelo, C. (2017). University students' self-regulated learning using digital technologies. *International Journal of Educational Technology in Higher Education*, *14*(48), 2017. doi:10.118641239-017-0076-8

Zhao, Y., & Lai, C. (2007). Technology and second language learning: Promises and problems. In L. L. Parker (Ed.), *Technology-mediated learning environments for young English learners: Connections in and out of school* (pp. 167–205). Mahwah, NJ: Lawrence Erlbaum Associates.

Zimmerman, B. J. (1989). A social cognitive view of self-regulated academic learning. *Journal of Educational Psychology*, *81*(3), 329–339. doi:10.1037/0022-0663.81.3.329

Zimmerman, B. J. (1995). Self-regulation involves more than metacognition: A social cognitive perspective. *Educational Psychologist*, *30*(4), 217–221. doi:10.120715326985ep3004_8

Zimmerman, B. J. (1998). Developing self-fulfilling cycles of academic regulation: An analysis of exemplary instructional model. In D. H. Schunk & B. J. Zimmerman (Eds.), *Self-regulated learning: From teaching to self-reflective practice* (pp. 1–19). New York: Guilford.

Zimmerman, B. J. (2000). Attaining self-regulation: A social cognitive perspective. In M. Boekaerts, P. R. Pintrich & M. Zeidner (Eds.), Handbook of self-regulation (pp. 451-502). San Diego, CA: Academic Press.

Zimmerman, B. J. (2002). Becoming a self-regulated learner: An overview. *Theory into Practice*, *41*(2), 64–70. doi:10.120715430421tip4102_2

Zimmerman, B. J., & Martinez-Pons, M. (1990). Student differences in self-regulated learning: Relating grade, sex, and giftedness to self-efficacy and strategy use. *Journal of Educational Psychology*, *82*(1), 51–59. doi:10.1037/0022-0663.82.1.51

Zimmerman, B. J., & Schunk, D. (2008). Motivation: An essential dimension of self-regulated learning. In B. J. Zimmerman & D. Schunk (Eds.), *Motivation and self-regulated learning: theory, research and applications* (pp. 1–30). New York: Lawrence Erlbaum Associates, Publishers.

ADDITIONAL READING

Benson, P. (2011). Language learning and teaching beyond the classroom: An introduction to the field. In P. Benson & H. Reinders (Eds.), *Beyond the language classroom* (pp. 7–16). Basingstoke: Palgrave Macmillan. doi:10.1057/9780230306790_2

Lai, C. (2013). A framework of developing self-directed technology use for language learning. *Language Learning & Technology*, *17*(2), 100–122. https://scholarspace.manoa.hawaii.edu/bitstream/10125/44326/1/17_02_lai.pdf

Lai, C., & Gu, M. Y. (2011). Self-regulated out-of-class language learning with technology. *Computer Assisted Language Learning*, *24*(4), 317–335. doi:10.1080/09588221.2011.568417

Lai, C., Zhu, W. M., & Gong, G. (2015). Understanding the quality of out-of-class English learning. *TESOL Quarterly*, *49*(2), 278–308. doi:10.1002/tesq.171

Pintrich, P. R. (2004). A conceptual framework for assessing motivation and self-regulated learning in college students. *Educational Psychology Review*, *16*(4), 385–407. doi:10.100710648-004-0006-x

Richards, J. C. (2015). The changing face of language learning: Learning beyond the classroom. *RELC Journal*, *46*(1), 5–22. doi:10.1177/0033688214561621

Yot-Domínguez, C., & Marcelo, C. (2017). University students' self-regulated learning using digital technologies. *International Journal of Educational Technology in Higher Education*, *14*(48), 2017. doi:10.118641239-017-0076-8

Zimmerman, B. J. (1989). A social cognitive view of self-regulated academic learning. *Journal of Educational Psychology*, *81*(3), 329–339. doi:10.1037/0022-0663.81.3.329

Zimmerman, B. J. (1995). Self-regulation involves more than metacognition: A social cognitive perspective. *Educational Psychologist*, *30*(4), 217–221. doi:10.120715326985ep3004_8

KEY TERMS AND DEFINITIONS

Out-of-Class Learning: Learning activities undertaken beyond the classroom.

Pre-Service Teacher: Undergraduate students enrolled in a teacher education program.

Self-Regulated Learning: Academic skills through which learners can take active roles in their own learning processes.

APPENDIX 1

Survey of Self-Regulated Learning with Technology at the University (SRLTU)

1. I comment on the information provided by my instructors during class using Twitter, Facebook, etc.
2. With my mobile device, I record the audio of my instructors' presentations.
3. While studying, I review my instructors' presentations, which were created using PowerPoint, Impress, Prezi, etc.
4. I share study materials and resources with my classmates using Dropbox, Facebook, Google+, etc.
5. While studying, I watch videos on YouTube, Vimeo, etc.
6. I listen to podcasts in which concepts and ideas related to course content are explained.
7. I record my own podcasts about course-related content to help me study.
8. I use specialized databases (Google Scholar, Proquest, Scopus, etc.) to find scholarly articles about course-related content.
9. I use online translation tools or mobile translation apps to better understand articles written in other languages about course-related material.
10. While studying, I create concept maps using specific software (CmapTools, MindManager, etc.) to help me understand the structure and relationships between concepts covered in my courses.
11. I complete assignments, in collaboration with my classmates, using tools such as wikis, Google Drive, Dropbox, etc.
12. I interact with my classmates outside the classroom setting using mobile apps (WhatsApp, Viber, Tango, Line, etc.) exchanging information, clarifying concepts, etc.
13. I communicate with my classmates using videoconference tools (Skype, Google Talk, FaceTime, TeamViewer, etc.) to clarify concepts or discuss topics related to the courses I am taking.
14. I have a blog where I give my point of view regarding course-related content.
15. I share images of in-class activities using Instagram, Flickr, Pinterest, Facebook, Twitter, or other similar platforms.
16. I use citation management tools such as Zotero, RefWorks, EndNote, and Mendeley when I have to write reports, essays, articles, etc.
17. I read news, information, and updates about course-related content that are shared on social media (Facebook, Twitter, Google+, etc.).

18. I am part of groups within social networks (LinkedIn, Facebook, Google+, etc.) that discuss and exchange information about topics related to the material covered in the courses that I am taking.
19. I use programs such as Google Calendar and Evernote to manage my schedule of academic activities.
20. I partake in academic activities such as conferences, webinars, or other open-access initiatives (MOOCs, OpenCourseWare, etc.) that cover course-related content.
21. I download materials related to the topics covered in class from my university's repository of learning tools or from open-access learning resource libraries.
22. When I prepare for a presentation, a practical exam, or other similar tasks, I take videos of myself using my mobile device.
23. I create sample tests online (ExamTime, Google Forms, etc.), and share them with my classmates when I prepare for exams.
24. I find self-assessment exercises online related to the material covered in class, and I do them to prepare for exams.
25. Before submitting assignments, I verify their originality using anti-plagiarism software.
26. I use social bookmarking services such as Diigo or Delicious to record and store information I find online.
27. I use RSS readers to get up-to-date information regarding topics that interest me.
28. When I prepare a presentation (PowerPoint, Impress, Prezi, etc.), I share it using services like SlideShare.
29. I digitize my notes and assignments, and convert them into e-books using online services like Scribd, Issuu, and Calaméo to facilitate their distribution.
30. I use mobile apps such as Kalkulilo (the scientific calculator), Whiteboard Lite (the collaborative drawing app), and iConversor (the currency convertor) to solve problems and complete course exercises.
31. I follow YouTube channels where videos related to course content are posted.
32. I consult Wikipedia or online dictionaries when I need to clarify a matter or concept related to the material covered in class.
33. I follow blogs written by experts who post about matters related to course content.

Chapter 7
EFL Learning Beyond the Wall With MALL:
College Students' Perceptions

Mohialdeen Alotumi
https://orcid.org/0000-0001-7325-5942
Sana'a University, Yemen

ABSTRACT

This chapter reports an exploratory study that examines Yemeni college students' perceptions of mobile-assisted language learning (MALL), using their mobile phones, in their informal learning of English as a Foreign Language (EFL). The sample population comprised 150 EFL college students. Data were obtained through a questionnaire based on technology acceptance model (TAM). The findings revealed that EFL college students had positive perceptions towards utilizing MALL for informal EFL learning. Most participants maintained positive perceptions on the two constructs of TAM: perceived usefulness and perceived ease of use. They believed that utilizing MALL furthered their EFL learning. Therefore, policymakers, curriculum designers, and educators are recommended to capitalize on using mobile phones for informal autonomous EFL learning in a way that can serve the objectives of formal EFL learning.

INTRODUCTION

Computer-assisted language learning (CALL) is progressively gaining momentum, thus, impacting many aspects of language teaching and learning (Beatty, 2010) as well as intercultural awareness and understanding (Schulze & Smith, 2017). In

DOI: 10.4018/978-1-7998-2116-8.ch007

Copyright © 2020, IGI Global. Copying or distributing in print or electronic forms without written permission of IGI Global is prohibited.

today's Information Age, students are considered digital natives (Prensky, 2001), presumably inclined and eager to take advantage, to the utmost, of the information and communication technology (ICT) in their learning in general and in their learning of English in particular. In Yemen, English as a foreign language (EFL) is formally taught as a school subject, starting from the 7th grade to the 12th grade—three to four periods of 45 minutes per week. At the university level, EFL is a requirement for all non-English majors as well as a major of study (Alotumi, 2018). In such an EFL context, students' exposure to English is feasible mostly in formal settings such as school, institute, or college. However, in informal settings, through CALL, EFL learners can avail themselves of the authentic and non-authentic English language materials available online or offline (Hsu, Hwang, & Chang, 2013; Kukulska-Hulme, 2009; Shinagawa, 2011; Sutherland, Robertson, & John, 2009).

With the recent technological developments that are characterized with mobility, connectivity, and affordability, a new form of CALL has come into existence, namely, mobile-assisted language learning (MALL). Nowadays, most learners do have high-end smart phones or other mobile devices that are powerful, efficient, and with many affordances that could facilitate and enrich informal learning beyond the classroom due to "the wealth of digital resources and informal learning opportunities available" (Pegrum, 2014, p. 26). Made viable through mobile devices, MALL associates itself with ubiquity of language learning as Kukulska-Hulme, Lee, and Norris (2017) assert, "the use of mobile devices also potentially extends learning beyond the classroom setting" p. (217). In other words, transcending the enclosed space and limited time of classroom is what makes MALL intriguing. In this regard, Burston (2014b) underscores, "MALL can make significant contributions to improving language learning, most particularly by increasing time spent on language acquisition out of class" (p. 17). However, such premise needs to be scrutinized across different contextual settings in general and informal EFL learning in particular, for the proliferation of mobile technology renders both opportunities and challenges. Besides, its benefits for language learning are not guaranteed (García Botero, Questier, & Zhu, 2018; Burston, 2014a; Godwin-Jones, 2017; Pegrum, 2014). Accordingly, students' readiness for MALL should be assessed first and foremost (Al-Emran, Mezhuyev, & Kamaludin, 2018).

Keeping abreast of technological developments that have made ubiquitous learning attainable, numerous studies investigated user acceptance of mobile learning in different postsecondary contexts (Aburub & Alnawas, 2019; Al-Emran et al., 2018; García Botero, Questier, Cincinnato, He, & Zhu, 2018). Moreover, several studies examined the acceptance and usage of MALL for informal English as a second language (ESL) learning (Wan Azli, Shah, & Mohamad, 2018; Mekhzoumi, Hamzah, & Krishnasamy, 2018; Soleimani, Ismail, & Mustaffa, 2014) and for informal EFL learning (Chung, Chen, & Kuo, 2015; Gamble, 2017).

Though user acceptance of MALL is essential and varies across the different educational contexts (García Botero, Questier, Cincinnato, et al., 2018), there has not been any study—to the best knowledge of the researcher—examining MALL acceptance for informal EFL learning with reference to the Yemeni context. Therefore, this study aims to look into the acceptance of MALL use for informal EFL learning by Yemeni EFL college students. Specifically, the study investigates the Yemeni EFL college students' perceptions on (a) the perceived usefulness (PU) of MALL for informal EFL learning, and (b) the perceived ease of use (PEoU) of MALL for informal EFL learning. The findings of this exploratory study could impart insights for policy-makers, curriculum designers, and educators on college students' acceptance and usage of MALL for informal EFL learning. In doing so, this study will be a momentum for capitalizing on MALL for formal EFL learning and for further investigation relating to ICT integration into EFL education.

LITERATURE REVIEW

Technology Acceptance Model

Derived from Ajzen and Fishbein's (1980) Theory of Reasoned Action (TRA), Technology Acceptance Model (TAM) was first proposed by Davis (1986) in his doctoral dissertation entitled, "A Technology Acceptance Model for Empirically Testing New End-User Information Systems: Theory and Results", then modified by Davis, Bagozzi, and Warshaw (1989), and later refined by Venkatesh and Davis (1996). TAM is "an adaptation of TRA specifically tailored for modelling user acceptance of information systems" (Davis et al., 1989, p. 985). It empirically presents a cogent explication for an individual's ICT acceptance and usage behavior. It is also a valid and reliable framework commonly used in literature during the past decade for predicting and explicating a person's behavioral intention towards ICT use (Rahimi & Pourshahbaz, 2019).

TAM postulates that an individual's behavioral intention to adopt ICT can be predicted by two constructs: *perceived usefulness (PU)* and *perceived ease of use (PEoU)* (see Figure). PU refers to "the degree to which a person believes that using a particular system would enhance his or her job performance", while PEoU is characterized as "the degree to which a person believes that using a particular system would be free of effort" (Davis, 1989, p. 320). Simply put, TAM theorizes that if ICT is regarded useful and uncomplicated, the individual will develop a stronger intention to put it to good use. Correspondingly, EFL college students' positive perceptions of MALL usefulness and complexity will enable them develop strong intentions to make use of it and therefore affect their actual use of MALL for EFL learning (Wan Azli et al., 2018; Chung et al., 2015; Gamble, 2017).

A number of studies have adopted TAM to inquire into student acceptance of mobile learning at college level in numerous settings. For instance, Aburub and Alnawas (2019) carried out a study, combining key components of TAM as well as Usage and Gratification approach (U & G), to develop a new model for predicting user adoption of mobile learning in higher education. They collected data on 820 students studying in different majors at 10 Jordanian universities. Employing structural equation model, their findings revealed that PEoU, PU, cognitive gratification, and hedonic gratification were significant predictors for user acceptance of utilizing mobile learning at tertiary education.

Al-Emran et al. (2018) conducted a systematic review and synthesis of TAM-adopting mobile learning research from 2006 to 2018. Applying the inclusion/exclusion criteria on 799 publications, they filtered out only 82 studies eligible for analysis. Their findings indicated that most TAM studies pertinent to mobile learning put emphasis on extending TAM with external variables and combining factors from other theories with TAM. They also revealed that the main TAM-germane research problem was user acceptance of mobile learning and the main research tool employed was questionnaire. Further, they indicated that most of TAM studies focusing on mobile learning were done in post-secondary settings.

Fagan (2019) extended the Unified Theory of Acceptance and Use of Technology Model (UTAUT), including TAM, with hedonic motivation and social influence to probe into the acceptance of m-learning by students in a tertiary education milieu. She administered a survey on a sample of 171 university students. The findings of partial least squares structural equation modelling showed that enjoyment and performance expectancy were significant predictors of the acceptance of m-learning

Figure 1. Final version of TAM (Adapted from Venkatesh & Davis, 1996, p. 453)

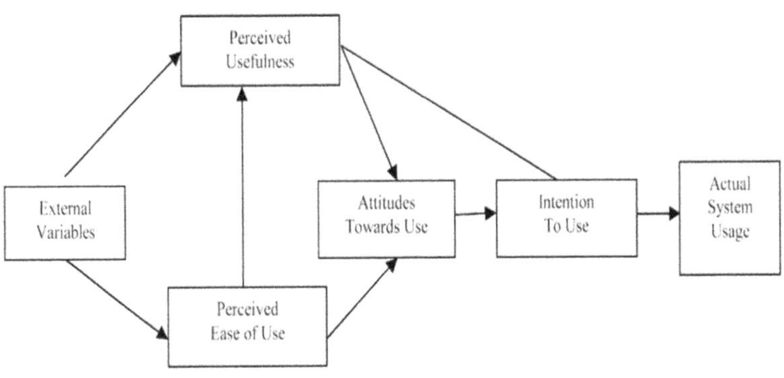

by college students. Overall, the students indicated they availed themselves of and enjoyed the use of iPads in doing their learning tasks.

García Botero, Questier, Cincinnato, et al. (2018) employed both TAM and UTAUT in assessing student adoption of MALL in higher education. They administered a questionnaire on a sample of 587 volunteering college students. Utilizing structural equation modelling, their findings demonstrated that the variables of facilitating conditions, performance expectancy, and social norms had an impact on participants' attitudes towards the use of MALL at college. They concluded that attitude was the most significant factor influencing students' behavioral intention, which affects their actual use of MALL.

Seliaman and Al-Turki (2012) put forward an extended TAM by including perceived innovativeness (PI) and ICT anxiety (IA). They used their proposed TAM to dissect the use of mobile phones and tablets by Saudi university students for doing course assignments, browsing the Internet to find course-related information, gaining access to course-related materials, and sharing information with classmates. Administering an extended-TAM questionnaire survey on a purposive sample of 60 male students from a computer and IT college, they found that though students' PI positively correlated with their PEoU, IA, and behavioral intention to use mobile learning, it had no significantly positive correlation with their PU of mobile learning.

MALL Acceptance for Tertiary ESL Learning

Recent research has adopted TAM to explore college students' acceptance of MALL for informal ESL learning in different educational settings. Wan Azli et al. (2018), for example, employed TAM to examine vocational college students' acceptance of using smart phones for their ESL learning. They administered a TAM-based questionnaire survey on a convenience sample of 100 undergraduate students studying in a vocational college in Malaysia. They found that participants held positive perceptions on both the PU and PEoU of MALL. They also indicated that employing MALL could bolster language learning as well as teaching.

Mekhzoumi et al. (2018) utilized a combination of TAM and UTAUT to look into students' adoption of MALL for their informal ESL learning. They used an acceptance and usage survey questionnaire on a sample of 348 undergraduates majoring in English and 326 undergraduates majoring in other subjects. They found that most participants had positive perceptions on the usage of MALL. They further demonstrated that student acceptance of MALL was determined by facilitating conditions, performance and effort expectancy, learning self-management, and social norms.

Soleimani et al. (2014) employed TAM to scrutinize ESL postgraduates' acceptance of MALL in Malaysia. They administered a TAM-based questionnaire survey on a convenience sample of 25 ESL postgraduates at a Malaysian university. In their findings, they revealed that the majority of respondents indicated positive perceptions on the PU and PEoU of MALL. They concluded that using mobile phone could render a convenient way for students to foster their ESL learning.

MALL Acceptance for Tertiary EFL Learning

In the context of EFL learning, Chung et al. (2015) combined TAM with compatibility and self-efficacy to examine EFL university students' acceptance of MALL-based resources designed for vocabulary learning. They administered a questionnaire on a convenience sample of 82 Taiwanese undergraduate students. In their findings, they verified that mobile devices' compatibility, self-efficacy, PEoU, and PU respectively had a significant impact on students' behavioral intention to utilize mobile devices in learning vocabulary. They concluded that compatibility is the most significant determiner of MALL use for vocabulary learning.

Gamble (2017) employed TAM to explore student acceptance of adopting Google Sites (GS), served as a Course Management System (CMS). He administered a questionnaire on a sample of 35 Japanese EFL college students. The findings revealed that most respondents reflected moderately positive perceptions towards the use of GS as a CMS. Furthermore, positive and significant relationships were found for all the constructs of PU, PEoU, attitude, and behavioral intentions. The study concluded that TAM could be beneficial framework to predict user acceptance of e-tools before implementation.

Hsu (2013) developed a questionnaire from literature review, including TAM, to explore EFL learners' perceptions of MALL at three Taiwanese universities. He administered a questionnaire on a convenience sample of 45 international postgraduate students studying EFL as a requirement in their graduate studies. He found a significant difference in the perceptions of the culturally different informants. He also indicated that there was a consensus of opinion among the respondents that MALL had a potential to render for EFL learning.

MALL Effect on ESL/EFL Learning

Numerous studies have looked into the benefits of using MALL for EFL/ESL formal and informal learning in various international contexts (Burston, 2014b; Cho, Lee, Joo, & Becker, 2018; Dehkordi & Taki, 2018; Fucekova & Metruk, 2018; Ko, 2019; Luis, 2018; Morgana, 2017; Wrigglesworth, 2019; Zhonggen, Ying, Zhichun, & Wentao, 2018). Cho et al. (2018) conducted an exhaustive review of MALL-related studies to enquire into the impact of MALL on student achievement in informal and formal ESL/EFL learning. They employed the inclusion/exclusion rules on 2869

MALL-related articles published in the period of 2005-2017. Out of 20 studies conducted in different contextual settings, they extracted 22 effect sizes. In their meta-analysis, they used six variables as potential moderators, namely, study source, study context, test type, level of school, ESL/EFL language learner, and skill of ESL/EFL language learning. They found that utilizing mobile phones or tablets had a moderate positive overall effect on the achievement of students in language acquisition and language learning. They concluded that such an effect proved that MALL could facilitate L2 learning.

Another comprehensive meta-analysis was carried out by Burston (2014b). He critically analyzed 575 publications germane to MALL, encompassing 347 empirical studies, published in the period of 1994-2012. He conducted his analysis focusing on validity. He reported no effect size in his study, claiming that the bulk of MALL research had design shortcomings for the opted five moderator variables, namely, speaking, writing, listening, reading, and vocabulary. His analysis indicated that only 15 studies, with focus on reading, listening and speaking, furnished evidence of MALL advantage and four vocabulary-focused studies showed no significant difference.

Lin and Lin (2019) focused on vocabulary in their thorough review of MALL-related research conducted from 2005 to 2018. Applying the criteria of inclusion/exclusion on 83 publications relating to MALL-based target language vocabulary learning, they examined 33 eligible studies for the effect of using MALL on target-language word retention. They found that MALL use had positively large impact on learning target-language vocabulary. They also indicated that the mode of SMS/MMS outdid the mode of mobile application with respect to the benefits of mobile target-language word retention. They concluded with a disclaimer that such findings should be considered suggestive and not definitive due to the small sample size and design flaws in the main studies, when compared with reference to the selected moderator variables including study context, intervention time, autonomy of tasks, and outcome measurement of target-language vocabulary.

METHODOLOGY

Considering that it is the first of its kind to inquire into college students' acceptance of MALL for informal EFL learning in Yemen, the study reported in this chapter is exploratory in nature. According to Cohen, Manion, and Morrison (2018), an exploratory study investigates "the interpretations of the situation made by the participants themselves, to understand their attitudes, behaviors and interactions" (p. 20).

Sample and Setting

The study sample consisted of 150 college students, 19-24 years old. They majored in English at the English Department of the Languages Faculty, Sana'a University, Yemen. All of them were in the second semester of their first academic year of 2018-2019. The selection was based on the notion that Yemeni students in the first year of their English major study courses aim to improve their English proficiency. A convenience sampling was employed in this study since the study targeted college level-one students, majoring in English at the languages Faculty of Sana'a University, and since the respondents were selected according to their willingness to partake in the study. According to Cohen et al. (2018), a convenience sample is based on non-probability sampling, which "derives from the researcher targeting a particular group" (p. 217).

Instruments

The study utilized a questionnaire survey, a modified version of Davis' (1993) questionnaire, for collecting data from the participants (see Appendix). The questionnaire is convenient for numerical data collection from a large number of respondents with less time and effort. It is also straightforward to dissect according to Cohen et al. Cohen et al. (2018), "The questionnaire is a widely used and useful instrument for collecting survey information, providing structured, often numerical data, able to be administered without the presence of the researcher and often comparatively straightforward to analyze" (p. 471). The questionnaire of this study is made of three sections.

Section A encompassed three items that sought information on the participants' demographic information. Section B comprised a scale of 10 statements (items 1-10), measuring students' perceptions on the PU of MALL for informal EFL learning. Section C consisted of a 10-item scale (items 11-20), assessing students' perceptions on the PEoU of MALL for informal EFL learning. Respondents replied to each statement of sections B and C in the light of a 7-point Likert-type scale with responses ranging from *strongly disagree =1*, to *disagree =2*, to *slightly disagree =3*, to *don't know = 4*, to *slightly agree = 5*, to *agree = 6*, to *strongly agree = 7*. Though Davis' (1993) questionnaire had high reliability and validity scores, the modified version was piloted to ensure its reliability and validity before administering it on the study sample.

Data Collection and Analysis Procedures

After piloting the questionnaire survey, adding the necessary modifications to the instrument, the researcher coordinated with level-one EFL student representatives to gather their classmates who were willing to participate in the study in a lecture hall for administering the questionnaire. Indeed, 150 level-one EFL college students were seated in a lecture hall waiting for the researcher. Upon arrival, and before distributing the questionnaire, the researcher introduced the aim and scope of the study and reflected his appreciation of students' time and effort to be part of this study. Afterwards, the questionnaire copies were handed over to the attending participants, who were given 60 minutes to respond to the questionnaire items. As the respondents were willing to partake in the study, the valid response rate of returned questionnaires was 100%.

The quantitative data obtained by using the questionnaire were analyzed through the Statistical Package for the Social Sciences (SPSS v. 24). The analysis employed was descriptive statistics (frequency, percentage, mean, and standard deviation) to obtain participants' perceptions on the PU and PEoU of MALL for informal EFL learning.

FINDINGS

Respondents' Demographic Information

The purpose of respondents' demographic information is to give a clear idea about the background of the study participants in terms of age, gender, and grade point average (GPA). Table 1 shows the percentage and frequency of respondents' age, gender, and GPA. As regards age, all participants were in the same year group of 19-24 years old. In terms of gender, the majority of respondents in this study were females (64%). Concerning GPA, almost all respondents got a GPA of above 'acceptable': 42% of the respondents scored 'very good'; 30.7% scored 'excellent'; and 24.7% scored 'good' in the first semester of their first year. Students' GPA indicated a high proficiency in English.

Students' Perceptions on the PU of MALL

Participants were directed to respond to 10 statements (items 1-10) as per a 7-point Likert-type scale, measuring their perceptions on the PU of MALL for informal EFL learning. Responses were ranked from 1 = strongly disagree (minimum value) to 7 = strongly agree (maximum value). Higher scores pointed to more positive

Table 1. Demographic background of respondents

Age			Gender		GPA				
19-24 years	25-29 years	30+ years	Male	Female	Exce.	Very good	Good	Acceptable	Failing
100.0 (150)	0 (0)	0 (0)	36.0 (54)	64.0 (96)	30.7 (46)	42.0 (63)	24.7 (37)	2.7 (4)	0 (0)

perceptions while lower scores indicated less positive perceptions towards the PU of MALL for informal EFL learning. Descriptive statistics for the participants' responses on the PU scale are presented as percentage and frequency (see Table 2) as well as mean and standard deviation (see Table 3).

As shown in Table 2, regarding item 1, 82.7% of the participants agreed or strongly agreed that using mobile phone enhanced the quality of the English-learning tasks they do. For item 2, 82% of the respondents agreed or strongly agreed that mobile phone gave them greater control over the English-learning tasks they do. Responding to item 3, 78% of the participants agreed or strongly agreed that mobile phone enabled them to fulfil the English-learning tasks faster. Concerning item 4, almost 75% of the respondents agreed or strongly agreed that mobile phone supported critical aspects of the English tasks they do. As per item 5, slightly more than 73% of the participants agreed or strongly agreed that using mobile phone stepped up their productivity while doing the English-learning tasks.

In addition, with reference to item 6, most respondents (78%) agreed or strongly agreed that using mobile phone improved their performance in doing the English-learning tasks. Regarding item 7, approximately 75% of respondents agreed or strongly agreed that using mobile phone allowed them to carry out more English-learning tasks than doing mobile-free tasks. About item 8, slightly more than 75% of the participants agreed or strongly agreed that using mobile phone enhanced their effectiveness on completing the English-learning tasks. About item 9, slightly more than 83% of the participants agreed or strongly agreed that using mobile phone made it easier for them to do English-learning tasks. With respect to the last item of the PU scale (item 10), almost 77% of the respondents agreed or strongly agreed in their overall assessment that mobile phone was useful in learning English outside classroom.

Table 3 points out the mean score and standard deviation for each item on the PU scale. It demonstrates that participants' highest mean score on the PU scale was 6.17 ($SD = 1.24$) and the lowest mean score was 5.61 ($SD = 1.44$). The overall mean score on the PU scale was 5.78 ($SD = 1.86$), indicating that EFL college students hold positive perceptions towards the PU of using MALL for informal EFL learning.

Table 2. Percentage and frequency of students' perceptions on the PU of MALL

No.	Item	Percentage (Frequency)						
		STD	D	SD	DK	SA	A	STA
1.	Using mobile phone enhances the quality of the English-learning tasks I do.	1.3 (2)	5.3 (8)	2.0 (3)	2.7 (4)	6.0 (9)	44.7 (67)	38.0 (57)
2.	Using mobile phone gives me greater control over the English-learning tasks I do.	1.3 (2)	4.0 (6)	2.0 (3)	2.7 (4)	8.0 (12)	48.0 (72)	34.0 (51)
3.	Mobile phone enables me to fulfil the English-learning tasks faster.	3.3 (5)	4.0 (6)	6..0 (9)	2.7 (4)	6.0 (9)	42.0 (63)	36.0 (54)
4.	Mobile phone supports critical aspects of the English-learning tasks I do.	2.0 (3)	3.3 (5)	5.3 (8)	6.9 (9)	8.7 (13)	52.0 (78)	22.7 (34)
5.	Using mobile phone steps up my productivity while doing the English-learning tasks.	2.0 (3)	4.7 (7)	3.3 (5)	4.0 (6)	12.7 (19)	44.0 (66)	29.3 (44)
6.	Using mobile phone improves my performance in doing the English-learning tasks.	1.3 (2)	3.3 (5)	6.0 (9)	4.7 (7)	6.7 (10)	48.0 (72)	30.0 (45)
7.	Using mobile phone allows me to carry out more English-learning tasks than doing mobile-free tasks	2.7 (4)	4.7 (7)	6.0 (9)	2.7 (4)	9.3 (14)	46.7 (70)	28.0 (42)
8.	Using mobile phone enhances my effectiveness on completing the English-learning tasks.	1.3 (2)	6.7 (10)	5.3 (8)	4.0 (6)	7.3 (11)	50.0 (75)	25.3 (38)
9.	Using mobile phone makes it easier for me to do English-learning tasks.	0.7 (1)	2.0 (3)	3.3 (5)	1.3 (2)	9.3 (14)	31.3 (47)	52.0 (78)
10.	Overall, I find the mobile phone useful in learning English outside the classroom.	2.9 (4)	6.0 (9)	5.3 (8)	2.0 (3)	7.3 (11)	39.3 (59)	37.3 (56)

Note. STD = Strongly disagree; D = Disagree; SD = Slightly disagree; DK = Don't know; SA = Slightly agree; A = Agree; STA = Strongly agree.

Students' Perceptions on the PEoU of MALL

Participants were instructed to respond to 10 statements (items 11-20) as per a 7-point Likert-type scale, assessing their perceptions on the PEoU of MALL for informal EFL learning. Items 12, 14, 16, 18, and 20 were stated positively while items 11, 13, 15, 17, and 19 were stated negatively. Responses to the scale were ranged from 1 = strongly disagree (minimum value) to 7 = strongly agree (maximum value). Higher scores made for more positive perceptions while lower scores signaled less positive perceptions towards the PEoU of MALL for informal EFL learning. Descriptive statistics for the participants' responses on the PEoU scale are presented as percentage and frequency (see Table 4) as well as mean and standard deviation (see Table 5).

As displayed in Table 4, 72% of the respondents disagreed or strongly disagreed on item 11 that they found the mobile phone troublesome to use for English learning while 75.4% of them agreed or strongly agreed on item 12 that learning to handle

Table 3. Mean and standard deviations of students' perceptions on the PU scale

No.	Item	M	SD
1.	Using mobile phone enhances the quality of the English-learning tasks I do.	5.96	1.30
2.	Using mobile phone gives me greater control over the English tasks I do.	5.94	1.24
3.	Mobile phone enables me to fulfil the English-learning tasks faster.	5.71	1.62
4.	Mobile phone supports critical aspects of the English-learning tasks I do.	5.61	1.44
5.	Using mobile phone steps up my productivity while doing the English tasks.	5.70	1.40
6.	Using mobile phone improves my performance in doing the English tasks.	5.72	1.45
7.	Using mobile phone allows me to carry out more English-learning tasks than doing mobile-free tasks.	5.62	1.55
8.	Using mobile phone enhances my effectiveness on completing the English tasks.	5.61	1.46
9.	Using mobile phone makes it easier for me to do English-learning tasks.	6.17	1.24
10.	Overall, I find the mobile phone useful in learning English outside the classroom.	5.73	1.59

the mobile phone was trouble-free for them. Furthermore, approximately 73% of the respondents disagreed or strongly disagreed on item 13 that interacting via the mobile phone was often frustrating while slightly more than 67% of them agreed or strongly agreed on item 14 that they found it easy to get the mobile phone to perform what they needed it to do. Additionally, 76% of the respondents disagreed or strongly disagreed on item 15 that the mobile phone was difficult and hard to interact with while slightly more than 75% of them agreed or strongly agreed on item 16 that it was easy for them to think of how to do their English-learning task with the mobile phone.

With regard to item 17, the majority of participants (76%) disagreed or strongly disagreed that interacting with the mobile phone took up a lot of thinking. Conversely, on item 18, 70% of the respondents agreed or strongly agreed that their interaction using the mobile phone was plain and comprehensible. About item 19, almost 69% of the respondents disagreed or strongly disagreed that they found it required a lot of effort to be adept at using mobile phone. However, on the last item of the PEoU scale (item 20), most of the participants (72%) agreed or strongly agreed in their overall assessment that they found the mobile phone uncomplicated to utilize.

Table 5 shows the mean score and standard deviation for the items on the PEoU scale after altering the polarity of all negative items. It shows that participants' highest mean score responses on the PEoU scale was 6.11 ($SD = 0.79$) for finding mobile phone convenient to use (item 11) and the lowest mean score was 5.54 ($SD = 1.46$) for being easy to get the mobile phone to perform what they need it to do (item 14). The overall mean score on the PEoU scale was 5.85 ($SD = 1.09$), revealing

Table 4. Percentage and frequency of students' perceptions on the PEoU of MALL

No.	Items	Percentage (Frequency)						
		STD	D	SD	DK	SA	A	STA
11.	I find the mobile phone troublesome to use for English learning.	30.0 (45)	6.0 (9)	42.0 (63)	3.3 (5)	10.0 (15)	4.7 (7)	4.0 (6)
12.	Learning to handle the mobile phone is trouble-free for me.	2.0 (3)	7.3 (11)	4.0 (6)	1.3 (2)	10.0 (15)	44.7 (67)	30.7 (46)
13.	Interacting via the mobile phone is often frustrating.	21.3 (32)	12.0 (18)	51.3 (77)	3.3 (5)	5.3 (8)	4.0 (6)	2.7 (4)
14.	I find it easy to get the mobile phone to perform what I need it to do.	2.0 (3)	6.7 (10)	4.7 (7)	1.3 (2)	18.0 (27)	44.0 (66)	23.3 (35)
15.	The mobile phone is difficult and hard to interact with.	25.3 (38)	8.0 (12)	50.7 (76)	2.7 (4)	7.3 (11)	4.7 (7)	1.3 (2)
16.	It is easy for me to think of how to do my English-learning task with the mobile phone.	0.7 (1)	8.7 (13)	5.3 (8)	1.3 (2)	8.7 (13)	56.0 (84)	19.3 (29)
17.	Interacting with the mobile phone takes up a lot of thinking.	23.3 (35)	4.7 (7)	52.7 (79)	1.3 (2)	11.3 17	4.0 (6)	2.7 (4)
18.	My interaction using the mobile phone is plain and comprehensible.	0.7 (1)	10.7 (16)	4.7 (7)	2.0 (3)	12.0 (18)	41.3 (62)	28.7 (43)
19.	I find it requires a lot of effort to be adept at using mobile phone.	26.7 (40)	16.7 (25)	42.0 (63)	2.7 (4)	6.7 (10)	4.0 (6)	1.3 (2)
20.	Overall, I find the mobile phone uncomplicated to utilize.	1.3 (2)	6.7 (10)	2.7 (4)	0.7 (1)	16.7 (25)	32.7 (49)	39.3 (59)

Note. STD = Strongly disagree; D = Disagree; SD = Slightly disagree; DK = Don't know; SA = Slightly agree; A = Agree; STA = Strongly agree.

that EFL college students had generally positive perceptions toward the PEoU of using MALL for the informal EFL learning.

DISCUSSION

The results from the study survey evince that Yemeni EFL college students have positive perceptions towards using MALL for informal EFL learning with reference to both TAM constructs, namely, PU and PEoU. This finding is consistent with those of Chung et al. (2015) and Gamble (2017), revealing that EFL college students' PU and PEoU of mobile devices are significant in determining their usage of MALL.

The results reveal that most respondents hold positive perceptions on all the aspects of MALL PU. Such finding explains that using MALL for EFL learning is fraught with merits that students could avail themselves while learning EFL. According to Wrigglesworth (2019), EFL students perceive mobile phones as useful and therefore, they actively utilize the affordances of their mobile phones to help them in doing their EFL learning tasks. Consequently, students' positive perceptions on the usefulness of mobile phones could be viewed as an incentive for paving the way for the integration of MALL in formal EFL learning in the Yemeni universities.

In addition, the findings indicate that most EFL college students have positive perceptions on the utilization of MALL for informal EFL learning. This gives evidence that students perceive the use of MALL as instrumental in furthering their EFL learning. According to Ko (2019), college students generally find MALL familiar and convenient to use in their EFL learning because mobile phones are an integral part of their lives. In this regard, Dehkordi and Taki (2018) suggested that MALL can lend itself well in both extensive and intensive EFL learning, inside and outside the classroom. Thus, capitalizing on the untapped potential of mobile devices in general and smart phones in particular could be achieved through fostering MALL for both formal and informal EFL learning.

Table 5. Mean scores and standard deviations on the PEoU of MALL after reversing polarity

No.	Item	M	SD
11.	*I find the mobile phone troublesome to use for English learning.	6.11	0.790
12.	Learning to handle the mobile phone is trouble-free for me.	5.70	1.478
13.	*Interacting via the mobile phone is often frustrating.	6.00	0.742
14.	I find it easy to get the mobile phone to perform what I need it to do.	5.54	1.455
15.	*The mobile phone is difficult and hard to interact with.	6.06	0.726
16.	It is easy for me to think of how to do my English-learning task with the mobile phone.	5.57	1.402
17.	*Interacting with the mobile phone takes up a lot of thinking.	6.07	0.686
18.	My interaction using the mobile phone is plain and comprehensible.	5.59	1.485
19.	*I find it requires a lot of effort to be adept at using mobile phone.	5.99	0.790
20.	Overall, I find the mobile phone uncomplicated to utilize.	5.84	1.391

Note. *negative items were reversed before scoring.

CONCLUSION

This chapter highlighted an exploratory study on college students' perceptions of MALL acceptance for informal EFL learning. The findings demonstrate that EFL college students have positive perceptions on both the PU and PEoU of MALL for EFL learning. The study has two specific implications. First, MALL has a solid potential that can be put to good use. Second, EFL college students are ready for utilizing MALL in their EFL learning. Considering the potential of MALL and students' acceptance of MALL, educators are recommended to integrate MALL into their EFL instruction through allowing, guiding and encouraging learners to harness their mobile phones in a way to serve the pedagogical objectives of EFL lessons. Such integration can gain momentum through students' positive perceptions towards the usage of MALL for autonomous EFL learning. Moreover, policy-makers and curriculum designers should tap into the affordances of mobile devices, accommodating the needs of students in today's digital era, by including learning tasks that can be facilitated through students' use of the mobile phone. There are, however, two specific limitations to this study that may affect the generalizability of its findings to other contexts. First, the sampling procedure adopted was convenience sampling. Second, the sample size was only 150 participants. A similar study employing a larger random sample is thus recommended for further investigation.

ACKNOWLEDGMENT

The author is very thankful to level-one EFL students of the Faculty of Languages at Sana'a University for dedicating time and effort to participate in this study. He is also appreciative of the coordination done by their representatives.

REFERENCES

Aburub, F., & Alnawas, I. (2019). A new integrated model to explore factors that influence adoption of mobile learning in higher education: An empirical investigation. *Education and Information Technologies*, *24*(3), 2145–2158. doi:10.100710639-019-09862-x

Ajzen, I., & Fishbein, M. (1980). *Understanding attitudes and predicting social behaviour*. Englewood Cliffs, NJ: Prentice-Hall, Inc.

Al-Emran, M., Mezhuyev, V., & Kamaludin, A. (2018). Technology acceptance model in m-learning context: A systematic review. *Computers & Education, 125*, 389–412. doi:10.1016/j.compedu.2018.06.008

Alotumi, M. (2018). The effect of CALL-based instruction on students' score attainment on the TOEFL iBT in a Yemeni context. *International Journal of Computer-Assisted Language Learning and Teaching*, *8*(1), 50–64. doi:10.4018/IJCALLT.2018010104

Beatty, K. (2010). *Teaching and researching computer-assisted language learning* (2nd ed.). London: Pearson Education Limited.

Burston, J. (2014a). MALL: The pedagogical challenges. *Computer Assisted Language Learning*, *27*(4), 344–357. doi:10.1080/09588221.2014.914539

Burston, J. (2014b). Twenty years of MALL project implementation: A meta-analysis of learning outcomes. *ReCALL*, *27*(1), 4–20. doi:10.1017/S0958344014000159

Cho, K., Lee, S., Joo, M.-H., & Becker, B. (2018). The effects of using mobile devices on student achievement in language learning: A meta-analysis. *Education in Science*, *8*(3), 105. doi:10.3390/educsci8030105

Chung, H.-H., Chen, S.-C., & Kuo, M.-H. (2015). A study of EFL college students' acceptance of mobile learning. *Procedia: Social and Behavioral Sciences*, *176*, 333–339. doi:10.1016/j.sbspro.2015.01.479

Cohen, L., Manion, L., & Morrison, K. (2018). *Research methods in education* (8th ed.). London: Routledge.

Davis, F. D. (1986). *A technology acceptance model for empirically testing new end-user information systems: theory and results* (Doctoral dissertation). Retrieved from http://hdl.handle.net/1721.1/15192

Davis, F. D. (1989). Perceived usefulness, perceived ease of use, and user acceptance of information technology. *Management Information Systems Quarterly*, *13*(3), 319–340. doi:10.2307/249008

Davis, F. D. (1993). User acceptance of information technology: System characteristics, user perceptions and behavioral impacts. *International Journal of Man-Machine Studies*, *38*(3), 475–487. doi:10.1006/imms.1993.1022

Davis, F. D., Bagozzi, R. P., & Warshaw, P. R. (1989). User acceptance of computer technology: A comparison of two theoretical models. *Management Science*, *35*(8), 982–1003. doi:10.1287/mnsc.35.8.982

Dehkordi, M. E., & Taki, S. (2018). Iranian male and female EFL learners' perceptions toward the use of mobile assisted language learning. *Journal of Applied Linguistics and Language Research*, *5*(3), 56–66.

Fagan, M. H. (2019). Factors influencing student acceptance of mobile learning in higher education. *Computers in the Schools*, *36*(2), 105–121. doi:10.1080/07380569.2019.1603051

Fucekova, M., & Metruk, R. (2018). Developing English skills by means of mobile applications. *Information Technologies and Learning Tools*, *66*(4), 173–185. doi:10.33407/itlt.v66i4.2376

Gamble, C. (2017). Exploring EFL university students' acceptance of E-learning using TAM. *Humanities Report*, *22*, 23–37.

García Botero, G., Questier, F., Cincinnato, S., He, T., & Zhu, C. (2018). Acceptance and usage of mobile assisted language learning by higher education students. *Journal of Computing in Higher Education*, *30*(3), 426–451. doi:10.100712528-018-9177-1

García Botero, G., Questier, F., & Zhu, C. (2018). Self-directed language learning in a mobile-assisted, out-of-class context: Do students walk the talk? *Computer Assisted Language Learning*, *32*(1-2), 71–97. doi:10.1080/09588221.2018.1485707

Godwin-Jones, R. (2017). Smartphones and language learning. *Language Learning & Technology*, *21*(2), 3–17. Retrieved from http://llt.msu.edu/issues/june2017/emerging.pdf

Hsu, C.-K., Hwang, G.-J., & Chang, C.-K. (2013). A personalized recommendation-based mobile learning approach to improving the reading performance of EFL students. *Computers & Education*, *63*, 327–336. doi:10.1016/j.compedu.2012.12.004

Hsu, L. (2013). English as a foreign language learners' perception of mobile assisted language learning: A cross-national study. *Computer Assisted Language Learning*, *26*(3), 197–213. doi:10.1080/09588221.2011.649485

Ko, M.-H. (2019). Students' reactions to using smartphones and social media for vocabulary feedback. *Computer Assisted Language Learning*, 1–25. doi:10.1080/09588221.2018.1541360

Kukulska-Hulme, A. (2009). Will mobile learning change language learning? *ReCALL*, *21*(2), 157–165. doi:10.1017/S0958344009000202

Kukulska-Hulme, A., Lee, H., & Norris, H. (2017). Mobile learning revolution: Implications for language pedagogy. In C. A. Chapelle & S. Sauro (Eds.), *The handbook of technology and second language teaching and learning* (pp. 217–233). Hoboken, NJ: Wiley. doi:10.1002/9781118914069.ch15

Lin, J.-J., & Lin, H. (2019). Mobile-assisted ESL/EFL vocabulary learning: A systematic review and meta-analysis. *Computer Assisted Language Learning*, 1–42. doi:10.1080/09588221.2018.1541359

Luis, A. R. (2018). Perceptions of the educational benefits of mobile devices in language teaching and learning. In A. Rocha, H. Adel, L. Reis, & S. Costanzo (Eds.), Trends and Advances in Information Systems and Technologies: Vol. 746. Advances in Intelligent Systems and Computing (pp. 1406–1415). Cham: Springer. doi:10.1007/978-3-319-77712-2_135

Mekhzoumi, O., Hamzah, M. H. b., & Krishnasamy, H. N. (2018). Determinants of mobile applications acceptance for English language learning in Universiti Utara Malaysia. *Journal of Advanced Research Design*, *51*(1), 1–13.

Morgana, V. (2017). *The iPad and the development of speaking and writing in the secondary EFL classroom* (Doctoral dissertation). Retrieved from http://oro.open.ac.uk/56480/

Pegrum, M. (2014). *Mobile learning: Languages, literacies and cultures*. London: Palgrave Macmillan. doi:10.1057/9781137309815

Prensky, M. (2001). Digital natives, digital immigrants. *On the Horizon*, *9*(5), 1–6. doi:10.1108/10748120110424816

Rahimi, M., & Pourshahbaz, S. (2019). *English as a foreign language teachers' TPACK: Emerging research and opportunities*. Hershey, PA: IGI Global. doi:10.4018/978-1-5225-6267-2

Schulze, M., & Smith, B. (2017). The changing habitat of CALL. *Calico Journal*, *34*(3), i-iv. doi:10.1558/cj.34527

Seliaman, M. E., & Al-Turki, M. S. (2012). Mobile learning adoption in Saudi Arabia. *International Journal of Computer, Electrical, Automation. Control and Information Engineering*, *6*(9), 1129–1131.

Shinagawa, S. (2011). Adapting the iphone for language teaching and learning. In F. Zhang (Ed.), *Computer-enhanced and mobile-assisted language learning: Emerging issues and trends* (pp. 188–201). Hershey, PA: IGI Global. doi:10.4018/978-1-61350-065-1.ch009

Soleimani, E., Ismail, K., & Mustaffa, R. (2014). The acceptance of mobile assisted language learning (MALL) among post graduate ESL students in UKM. *Procedia: Social and Behavioral Sciences*, *118*, 457–462. doi:10.1016/j.sbspro.2014.02.062

Sutherland, R., Robertson, S., & John, P. (2009). *Improving classroom learning with ICT*. New York, NY: Routledge.

Venkatesh, V., & Davis, F. D. (1996). A model of the antecedents of perceived ease of use: Development and test. *Decision Sciences*, *27*(3), 451–481. doi:10.1111/j.1540-5915.1996.tb01822.x

Wan Azli, W. U. A., Shah, P. M., & Mohamad, M. (2018). Perception on the usage of mobile assisted language learning (MALL) in English as a second language (ESL) learning among vocational college students. *Creative Education*, *09*(01), 84–98. doi:10.4236/ce.2018.91008

Wrigglesworth, J. (2019). Using smartphones to extend interaction beyond the EFL classroom. *Computer Assisted Language Learning*, 1–22. doi:10.1080/09588221.2019.1569067

Zhonggen, Y., Ying, Z., Zhichun, Y., & Wentao, C. (2018). Student satisfaction, learning outcomes, and cognitive loads with a mobile learning platform. *Computer Assisted Language Learning*, *32*(4), 323–341. doi:10.1080/09588221.2018.1517093

ADDITIONAL READING

Alqahtani, M., & Mohammad, H. (2015). Mobile applications' impact on student performance and satisfaction. *TOJET: The Turkish Online Journal of Educational Technology*, *14*(4), 101–112.

Bull, P. H., & Patterson, G. C. (2016). Strategies to promote pedagogical knowledge interplay with technology. In J. Keengwe & G. Onchwari (Eds.), *Handbook of research on active learning and the flipped classroom model in the digital age* (pp. 255–271). Hershey, PA: IGI Global; doi:10.4018/978-1-4666-9680-8.ch013

Cohen, A., & Ezra, O. (2018). Development of a contextualised MALL research framework based on L2 Chinese empirical study. *Computer Assisted Language Learning*, *31*(7), 764–789. doi:10.1080/09588221.2018.1449756

Dashtestani, R. (2013). Implementing mobile-assisted language learning (MALL) in an EFL context: Iranian EFL teachers' perspectives on challenges and affordances. *The JALT CALL Journal*, *9*(2), 149–168.

Duvall, M., Matranga, A., Foster, A., & Silverman, J. (2016). Mobile learning: Technology as mediator of personal and school experiences. *International Journal of Game-Based Learning*, *6*(1), 30–42. doi:10.4018/IJGBL.2016010103

Gupta, S., Taneja, S., & Kumar, N. (2015). Redefining the classroom: Integration of open and classroom learning in higher education. In E. McKay & J. Lenarcic (Eds.), *Macro-level learning through massive open online courses (MOOCs)* (pp. 168–182). Hershey, PA: IGI Global; doi:10.4018/978-1-4666-8324-2.ch010

Jin, S. h. (2014). Implementation of smartphone-based blended learning in an EFL undergraduate grammar course. *Multimedia-Assisted Language Learning*, *17*(4), 11–37. doi:10.15702/mall.2014.17.4.11

Lai, C., & Zheng, D. (2017). Self-directed use of mobile devices for language learning beyond the classroom. *ReCALL*, *30*(3), 299–318. doi:10.1017/S0958344017000258

Lee, J. S. (2017). Informal digital learning of English and second language vocabulary outcomes: Can quantity conquer quality? *British Journal of Educational Technology*. doi:10.1111/bjet.12599

Medárová, V. (2014). E-learning tools applied in teaching English at the university level. In T. Issa, P. Isaias, & P. Kommers (Eds.), *Multicultural awareness and technology in higher education: Global perspectives* (pp. 156–175). Hershey, PA: IGI Global; doi:10.4018/978-1-4666-5876-9.ch008

KEY TERMS AND DEFINITIONS

Behavioral Intention: A mental attitude to make a choice and then take action to carry it out.

English as a Foreign Language (EFL): English is mainly used inside the class with zero or limited exposure to it outside the classroom.

English as a Second Language (ESL): English is used inside the class as well as outside the classroom.

Formal Learning: Learning that occurs in a formal setting under teacher tutelage.

Informal Learning: Learning that takes place outside the classroom without teacher oversight.

Mobile-Assisted Language Learning (MALL): The use of mobile device (e.g., smart phone) to foster learner language proficiency.

Perceived Ease of Use: The degree to which something is considered uncomplicated to utilize.

Perceived Usefulness: A benefit detected by a person's instinct or inference.

Technology Acceptance: A favorable reception of technology that develops an intention to adopt it.

APPENDIX

MALL Questionnaire for EFL College Students

Section (A): Selected Student Characteristics

Instructions: Kindly respond to the following questions by putting a tick mark () in the brackets that correspond to your answers. Please respond to all the questions.

1. How old are you?
   19-24 (__) 25-29 (__) 30 or more (__)
2. What is your **gender**?
   Male (__) Female (__)
3. How was your **Grade Point Average** in the first semester?

a. Excellent (__)	b. Very Good (__)	c. Good (__)	d. Acceptable (__)	e. Failing (__)

Section (B): Usefulness of Mobile Phone Instructions: Kindly indicate your perception of mobile phone usefulness in EFL learning outside the classroom as stated in each one of the statements below by drawing a circle around the number that corresponds to your agreement or disagreement level with it. Please respond to all statements.		Strongly disagree	Disagree	Slightly disagree	Don't know	Slightly agree	Agree	Strongly agree
No.	Items							
1.	Using mobile phone enhances the quality of the English-learning tasks I do.	1	2	3	4	5	6	7
2.	Using mobile phone gives me greater control over the English-learning tasks I do.	1	2	3	4	5	6	7
3.	Mobile phone enables me to fulfil the English-learning tasks faster.	1	2	3	4	5	6	7
4.	Mobile phone supports critical aspects of the English-learning tasks I do.	1	2	3	4	5	6	7
5.	Using mobile phone steps up my productivity while doing the English-learning tasks.	1	2	3	4	5	6	7
6.	Using mobile phone improves my performance in doing the English-learning tasks.	1	2	3	4	5	6	7
7.	Using mobile phone allows me to carry out more English-learning tasks than ding mobile-free tasks.	1	2	3	4	5	6	7
8.	Using mobile phone enhances my effectiveness on completing the English-learning tasks.	1	2	3	4	5	6	7
9.	Using mobile phone makes it for them easier to do English-learning tasks.	1	2	3	4	5	6	7
10.	Overall, I find the mobile phone useful in learning English outside the classroom.	1	2	3	4	5	6	7

Section (C): Mobile Phone Ease of Use Instructions: Kindly indicate your perception of mobile phone ease of use in EFL learning outside the classroom as stated in each one of the statements below by drawing a circle around the number that corresponds to your agreement or disagreement level with it. Please respond to all statements.		Strongly disagree	Disagreed	Slightly disagree	Don't know	Slightly agree	Agree	Strongly agree
No.	Items							
11.	I find the mobile phone troublesome to use for English learning.	1	2	3	4	5	6	7
12.	Learning to handle the mobile phone is trouble-free for me.	1	2	3	4	5	6	7
13.	Interacting via the mobile phone is often frustrating.	1	2	3	4	5	6	7
14.	I find it easy to get the mobile phone to perform what I need it to do.	1	2	3	4	5	6	7
15.	The mobile phone is difficult and hard to interact with.	1	2	3	4	5	6	7
16.	It is easy for me to think of how to do my English-learning task with the mobile phone.	1	2	3	4	5	6	7
17.	Interacting with the mobile phone takes up a lot of thinking.	1	2	3	4	5	6	7
18.	My interaction using the mobile phone is plain and comprehensible.	1	2	3	4	5	6	7
19.	I find it requires a lot of effort to be adept at using mobile phone.	1	2	3	4	5	6	7
20.	Overall, I find the mobile phone uncomplicated to utilize.	1	2	3	4	5	6	7

Thank you indeed for taking the time to complete this survey! You have been a cherished contributor to this research study and your efforts are highly valued.

Chapter 8
Formalizing Informal CALL in Learning English Language Skills

Norizan Abdul Razak
Universiti Kebangsaan Malaysia, Malaysia

Amr Abdullatif Yassin
Ibb University, Yemen

Tg Nor Rizan Tg Mohamad Maasum
Universiti Kebangsaan Malaysia, Malaysia

ABSTRACT

This chapter aimed at exploring how to formalize informal CALL in learning English language skills. The focus of the study was on learners' ICT skills and learning strategies and CALL elements. The data were collected through semi-structured interviews with four learners. The findings showed that adult learners in this age of technology do not need any training to use CALL to study English language. Moreover, adult learners have acquired the learning strategies needed for studying English informally through CALL. In terms of CALL elements, the findings showed that there is a need to integrate the elements of behaviorist CALL, cognitive CALL, and constructivist CALL to make informal CALL more effective. Therefore, the focus should be directed to providing professionally-designed CALL tools that suffice the needs of the learners by those who have educational background about language teaching and learning. This will enhance informal CALL learning and make self-directed learning more efficient.

DOI: 10.4018/978-1-7998-2116-8.ch008

Copyright © 2020, IGI Global. Copying or distributing in print or electronic forms without written permission of IGI Global is prohibited.

INTRODUCTION

The wave of IT (Information Technology) has influenced every aspect of our life, and technology has become a must in the field of language teaching and learning (Bahrani & Sim, 2012). Informal CALL (Computer Assisted Language Learning) has become a part of everyday activities through which learners can improve their language. This makes informal CALL a main trend in the process of language learning, as students all over the world spend time to learn English through CALL informally. Most of the students use CALL to improve their language, especially language skills. They come across and can easily access different materials (Selwyn & Gorard, 2004). This informal CALL has attracted most of the learners (Hague & Logan, 2009; Livingstone, 2001). That is, there is little need to attend a class or a course in order to use CALL to improve your language. Informal CALL materials are available all the time, which makes it a chance for interested students to access at any time to learn English language. However, informal CALL still needs more attention to make it more useful. The main concern regarding informal CALL is that students usually use websites and materials that might not be authentic. Perhaps, novice practitioners who have no background about learning theories or learning pedagogy prepare these materials and websites. This might make informal CALL not that useful as it lacks the basics of language teaching and learning.

Therefore, the fact is still that most learners use informal CALL to learn English language on their own. This makes the investigation in formalizing informal CALL an essential issue. To do so, there is a need to focus on learners' skills and CALL elements needed to formalize informal CALL. According to Kessler (2007), most of the teachers in the worldwide context lack formal preparation to use CALL in teaching. Although teachers are absent in informal CALL, their role is still essential in the process of designing and developing reliable CALL tools. Thus, the focus of this paper is on learners' ICT skills and learning strategies, and CALL elements.

Statement of the Problem

Informal learning has attracted many adult learners currently (Eraut, 2004; Hague & Logan, 2009; Livingstone, 2001), and using technology in informal learning showed that it is an effective method to improve students' language (Bahrani & Sim, 2012). However, informal CALL has not received enough attention from researchers and educators in order to maximize up the benefit to those who learn language informally. Moreover, informal ICT learning has proved to be an effective tool to the degree that learners might even surpass their teachers due to their learning strategies that might approach bricoleur-style as students link practice to thinking. This allows them to arrange and re-arrange the materials through inner dialogue or shared dialogue, resulting in a better informal learning (Levinsen 2007; Levinsen & Sørensen, 2011). However, other researchers argued that using technology informally is far away from formal technology learning, and learners cannot reach the school learning perspective because students need the essential student-teacher relationship in the process of learning (Levinsen & Sørensen, 2011). Such contradiction needs more investigation on the strategies that might reduce the gap between formal and informal CALL to compensate the role of the teacher.

Besides, Kessler (2007) stated that most of the teachers in the field of language teaching and learning lack the essentials for creating and integrating CALL inside the classroom. The author stated that those teachers have not received formal training in the area of CALL, and they tend to improve their knowledge and experience informally. This leads to different questions concerning CALL tools and materials: Are the teachers capable of designing CALL tools to students? If not, who designed such CALL tools? What is their background about language teaching and learning? What are the pedagogical elements of such CALL tools, if any? Do the designers use an instructional model in the process developing CALL tools? These are essential questions because preparing CALL tools is not an easy task, which increases the question on the authenticity of the materials as well. The idea of the materials should be highlighted, especially in informal CALL, because learners might not have the experience to decide on the suitability and correctness of the materials provided in CALL. Such issues need investigation to come up with implications to formalize informal CALL. This idea is supported by Levinsen and Sørensen (2011) who stated that formalized informal ICT learning cannot be achieved with trivia materials.

Besides, making informal CALL more formalized guide learners, teachers and designers to provide informal learners with better learning opportunities. This has nothing to do with settings because those learners do not attend formal settings to learn language; otherwise, CALL will be formal. Nevertheless, informal CALL learners can be directed to get better language learning, especially self-directed informal CALL learners. Accordingly, this paper suggests that to formalize informal CALL, there is a need to focus on learners and CALL tools to reduce the gap between informal and formal CALL and, therefore, formalize informal CALL. Teachers and designers are not absent in this scope because CALL tools are prepared by them for learners. Hence, this chapter aims to answer the following questions:

1. What are the ICT skills that learners need to formalize informal CALL?
2. What are the learning strategies that learners use to study informally through CALL?
3. What are the CALL elements that learners need to formalize informal CALL?

LITERATURE REVIEW

Formal, Non-Formal and Informal Learning

The difference between informal and formal learning has been of great interest to many researchers, and the debate goes on especially in the modern era when technology has become an essential part of the learning process. The difference between formal and informal language learning has been defined in several ways and by several authors. An earlier definition of informal learning was given by Coombs and Ahmed (1974) that informal learning refers to the process of gathering skills, knowledge, attitude as well as insights through the exposure to environment in an informal setting. This informal setting can be the home, the workplace or even the environment through reading books, listening to audios or watching TV. Therefore, this definition makes it clear that informal learning occurs in an informal setting which makes the process of learning unorganized and unsystematic. Lightbown and Spada (2001) give another distinction between informal and formal learning of a second or foreign language. The authors argued that informal learning is related to learning setting in which learners are exposed to the target language at work or at home or at any social interaction. However, formal learning occurs when the target language is taught to a group of second of foreign language learners in a formal setting such as a school or a university. This definition is related to the two terms learning and acquisition. That is, according to this argument formal learning refers to the process of language learning which occurs in any institute, while informal language learning refers to acquisition in which students acquire the target language through exposure and not by study. This is also supported by the argument of Lightbown and Spada (2001) that informal learning can occur in any place as the learners can surf the internet, watch video or listen to music to improve their language. This helps learners to improve their language indirectly as they can try to understand a movie or a song, or read the subtitles to understand what they are watching (Pemberton, Fallahkhair & Mosthoff, 2004). This makes it different from formal learning as the learners study the language structurally and purposefully in a classroom (Marsick & Watkins 1990). This is also supported by other scholars who see that the difference between formal learning and informal learning lies in the idea that formal learning is classroom-based and structured; whereas, informal learning happens outside the classroom and it is unstructured (Lightbown & Spada, 2001; Marsick & Watkins, 1990; Bahrani & Sim, 2012; Bahrani, Sim & Nekoueizadeh, 2014).

Moreover, some scholars have provided a further classification as they distinguished between formal, non-formal and informal learning (Merriam, Caffarella & Baumgartner, 2007). Non-formal learning is defined as "any organized, systematic, educational activity carried on outside the framework of the formal system to provide selected types of learning to particular subgroups in the population, adults as well as children" (Coombs & Ahmed, 1974, p. 8 as cited in Intarat, Chanchalor, & Murphy, 2017). Also, non-formal learning might have a simplified curriculum, life-long learning objectives and outcomes, less-qualified teachers, and flexible attendance (Singh, 2009; Intarat et al., 2017). This is different from informal learning which is led by the individual needs and his curiosity to learn rather than having a prescribed curriculum (Schwier, 2010). Also, informal learning does not require any kind of attendance and the teacher might not be involved in the learning process, while non-formal learning might be sponsored by an organization or the government of the country.

Besides, informal learning in the current age has been classified into incidental learning which happens with anyone in everyday activities of surfing the Internet. However, intentional informal learning – known also as self-directed learning – happens when the learner has a goal to achieve and starts to locate the sources to achieve this learning goal without the availability of an instructor or a teacher. This leads to the conclusion that formal learning has formal setting and structured materials; however, informal learning does have a formal learning setting and the materials are almost unstructured. Although non-formal learning has been discussed in different contexts, the distinction between formal and non-formal is still not clear because both have structured materials and non-formal learning might have a formal setting just like formal learning. Hence, it is better to classify non-formal learning under formal learning because non-formal learning is systematic as it provides the learners with structured materials to achieve a learning goals, especially that non-formal learning is almost supported by organizations or governments as a form of life-long learning.

CALL

The definitions of CALL have been on a constant change. Gunduz (2005) indicated that the computers were first used in the field of education in the 1960s even though they have been in hands with people since the first half of the 20th century. Also, Yang (2010) states that the use of computers in the process of learning started in the 1960s with the beginning of the emergence of computers. However, it was not widely spread until the end of the 20th century with the emergence of the personal computers. Computers were spread in schools and educational institutions in the 1980s, and software programs were spread in the market and became available for people during the 1980s as well (Gunduz, 2005).

The 21st century is considered the age of information technology in which technology plays a vital role especially in the field of education. One of the aspects of technology is the Internet which has influenced even the education approaches as the students can communicate with each other simultaneously (Gunduz, 2005). The wide spread of technology has empowered teachers and learners with effective tools for language learning which became known as CALL. The term CALL is related to methodology as it links between computer assisted materials and their application in the process of language teaching and learning (Levy, 1997). CALL is a form of computer-based learning. It is a method of learning as the materials in CALL are used to facilitate the process of learning, and it aims to make the process of learning student-centred. The philosophy of CALL is that students should study on their own using structured and/or unstructured lessons (Okonkwo, 2011). One of the definitions of CALL is given by Levy (1997) who defined it as "the search for and study of applications of the computer in language teaching and learning" (p. 1). This definition has a broad meaning as it includes any form of language learning using computers. This broad definition encapsulates different terms that were common in the 1980s and 1990s, including CAI (Computer-Assisted Instruction), CALI (Computer-Assisted Language Instruction) and TELL (Technology-Enhanced Language Learning) (Tafoli & Golshan 2014).

From another perspective, Beatty (2003) defined CALL as "any process in which a learner uses a computer and, as a result, improves his or her language" (p. 7). This definition highlights two components; the first one is computer and the second one is language improvement. The word computer does not refer to the laptop and desktop computers only as other devices were included at the beginning such as DVDs, PADs and MP3. In addition, the improvement might be in the language skills or other factors such as effectiveness, efficiency, time, effort, easiness and motivation. Besides, CALL definition is not limited to students only as teachers are also included since CALL provides a lot of assistance to teachers such as keeping records, evaluation and assessment as well as professional development and material development. Previously, using any form of technology was considered CALL; however, scholars have changed this perspective. According to Garrett (2009), the term CALL should link between theory, pedagogy and technology. This definition eliminates different forms of using computer from the circle of CALL.

Accordingly, the term CALL has been looked at from different perspective. Using technology and computers particularly is an important element in CALL; however, the later definitions clarify that using computers in the field of language teaching and learning might not be called CALL, unless it links between technology, theory and pedagogy. This concept should be taken into consideration in the attempts to formalize informal CALL.

Informal CALL

The area of formalizing informal CALL has not received enough attention from the side of researchers; however, one of the interesting studies in formalizing informal ICT is carried out by Levinsen and Sørensen (2011). This study used formal and informal interviews and observation to collect data, and it came out with interesting points concerning formalizing informal ICT learning. One of the findings of the study is that learners need to improve bricoleur skills that enables them to rethink of what they have learned in order to improve their learning. Moreover, the study stated that formalizing informal ICT learning might not be useful if the students are not provided with structured CALL materials, and this makes the role of the teacher essential besides technology.

The study by Pemberton, Fallahkhair and Masthoff (2004) tried to provide a theoretical framework in order to make language learning through interactive iTV formal. The study discussed that behaviourism and cognitivism are not suitable for formal interactive iTV, and the suitable theory is constructivism since it helps learners to construct their knowledge. However, the study by Bahrani and Sim (2012) showed that social interaction, which is an element of constructivism, is less effective than exposure to audio-visual materials. This study made a comparison between learning of speaking through exposure to audio-visual materials and social interaction. The findings of the study showed that exposure to audio-visual materials is more effective than social interaction. This supports that informal CALL is effective in teaching English language. The finding of the latter study is supported by the study of Trinder (2017) which showed that students prefer to use technology to learn, and they prefer structured materials.

Selwyn and Gorard (2004) stated that informal ICT increases the potential of learning among adults because of the notion of freeing as learners can learn at any time and in any place. They stated that ICT informal learning is the key to solve formal learning barriers stated in previous studies, including cultural, structural, personal, situational, institutional and dispositional barriers. The study of Selwyn and Gorard (2004) concluded that there is a need to put attention in utilizing technology in the field of education, because it is of great help for learners. This recommendation is supported by the study of Paredes et al. (2005) who showed that technology is of great asset for students to learn language in informal settings.

However, the study of Kessler (2007) directed the attention to the teacher preparation to use CALL, and the preparation of the teachers to use CALL in the process of language teaching and learning. The study showed that most EFL/ESL teachers receive informal preparation, and there is a need to focus on preparing teachers to use CALL in the process of language teaching and learning. He argued that it is vital to distinguish between formal and pedagogical use of technology, because the latter needs knowledge to create CALL materials which most of the teachers lack.

Moreover, informal ICT learning is of great asset to the students as it supports formal learning (Nawi, Yusoff, Ajmain, & Abbas, 2014; Ferguson, Faulkner, Whitelock & Sheehy, 2015). Also, Ferguson et al. (2015) stated that there is a need to improve learners' skills in order to make informal ICT learning more fruitful.

According to the discussion in this section, there is a need to focus on both teachers and learners in order to formalize informal CALL. However, the main role is still that of the teacher because he might be the teacher as well as the designer of the materials and CALL tools.

METHODOLOGY

This study adopted hermeneutical phenomenological approach. Phenomenological approach is a qualitative research design that describes the experience of a phenomenon for several individuals (Moustakas, 1994; Astalin, 2013; Hall, Chai & Albrecht, 2016). In this study, the experience is learning English language skills through informal CALL which includes using websites, online programs or any other forms of informal CALL. The purpose is to focus on the commonalities of the participants' experience in studying English informally through CALL. There are two types of phenomenological studies, namely hermeneutical and transcendental (Moustakas, 1994; Hall, Chai & Albrecht, 2016). The first one relies more on the researcher's interpretation for the participants' experiences, while the latter focuses more on the participants' description for their experience. This study used hermeneutical phenomenological approach since the researchers study the experience of the participants in using informal CALL to study English language skill. The interpretation of the researchers for the participants' experience is in the context of ICT skills, learning strategies and CALL elements. That is, the participants might not have a background about learning strategies or CALL element; however, it is the duty of the researchers to interpret their experience within the scope of the study. Moreover, the study employed semi-structured interviews method, which is suitable for the study for it allows generating new questions during the interviews. This helps to get an in-depth investigation for the participants' experience in studying English through informal CALL.

In terms of the participants, they were selected purposively according to the following criteria: (a) the learner should have informally studied the four skills (speaking, listening, reading and writing) through CALL, and (b) the learner should have enough experience in using CALL to study English language so that he can answer the interview questions according to his experience. Accordingly, most of the learners did not have enough experience in using CALL to study English language except four learners who agreed to participate in the study. Therefore, the limited number of the participants in this study is due to accessibility. That is, most of the requested learners to partake in the study stated that they have not focused on studying English using informal CALL. Therefore, they were not interviewed because their little experience will not serve the research purposes. The participants' background information is shown in Table 1 below.

In terms of data analysis, the researchers have analysed the data directly from the recordings. Halcomb and Davidson (2006) showed that this method is reliable because the researcher can make cross-checking directly from the recordings. According to Tessier (2012), analysing the data directly from recordings is the last evolution of data analysis as it saves time and effort, and it does not lead to the loss of context of the study. Moreover, Crichton and Child (2005) argued that coding directly from recordings has many advantages, including hearing the voice and intonation of the speaker and his/her reflections, reducing the loss of data because the researcher will not code from the transcripts which is another form of the interviews, keeping data true and fresh during the coding process, and reducing the risk of transcription errors, misrepresentation and loss of context.

Therefore, the researchers have taken verbal permission from participants at the beginning of all the interviews to use the data for research purposes. The collected data were analysed directly from the recordings in the form of thematic patterns to have a clear picture concerning self-directed CALL learners' ICT skills, learning strategies and CALL elements. The researchers listened to the recordings several times and took notes, which were developed into codes, and these codes were developed into themes at the end. In terms of reliability, Neal, Neal, VanDyke and Kornbluh (2015) stated that reliability of analysis from recordings can be achieved by allowing different experts to analyse the data. Accordingly, the analysis was carried out by one author and the other two authors have revised the analysis to ensure its reliability.

Table 1. Participants' background information

No.	Level of study	Major	Aim of using CALL
Participant 1	PhD	Structural engineering	Improve English language skills and take IELTS
Participant 2	Master	Civil engineering	Take IELTS
Participant 3	Master	Structural engineering and construction	Improve English language skills and take IELTS
Participant 4	Bachelor graduate	Information technology	Improve English language skills

FINDINGS

This section discusses the findings related to learners' skills needed to informalize learning of English language skills through CALL and the important elements that should be included in CALL tools. Learners' skills and CALL elements are discussed in the form of themes, which are used as a guide to formalize informal CALL.

Themes Related to Learners

The first theme is that self-learning using CALL is of great advantage to EFL learners. All the participants stated that they got the desired benefit. Some participants used CALL to study English language for IELTS, and this self-learning helped them to get almost the desired scores. They also stated that they could improve their English language through CALL because it saves time and provides learners with different resources to improve their skills. In this regard, participant 4 stated

What I do believe is that there is one difference when you study in institutes, sometimes you take the course as materials that you have to take and pass it. But, when you have self-study as self-study, you will learn as a language that you will need to enjoy, which motivates you more to enjoy the language itself more. So, self-study ... I do prefer self-study because it takes you higher than what the institute takes you usually.

Another theme is that learning English through CALL requires the basic skills for using computers only. That is, there are no special skills that are needed to study English through computer as any learner possesses the basic skills for using computer can easily study through CALL. Even in the process of searching for suitable materials, the participants stated that they have gained this skill of surfing the internet to find suitable materials with practice. This was expressed by all the participants in the study. For example, participant 2 stated

Anyone who can use a computer who knows to log into websites, create accounts in some websites aaa also the online lectures, lessons. Who has this ability can use it for English.

Also, participant 4 stated

What I believe is as long as you are educated, you will be able to use the computer and to learn [English] from the computer.

Another theme is that using computer is not limited to a specific slice of learners as students at different levels can use CALL to improve their English language. This is because there are many sources in the internet that suit all the students to improve their language. In this regard, participant 2 stated

Actually, it didn't matter. I mean, from a starter levels until the advanced of using a computer you can use to anywhere you go. So, as I mentioned previously, that some students like to use a computer and some like to use a paper. So, for using a computer is so helpful because you have the ability to access many websites to get many, uh, exercise from different sources. So, I think it's so much helpful.

The analysis showed that learners also have their own learning strategies to evaluate and to improve their English language. They usually reflect on their learning and come up with specific strategies in order to improve their language. In this regard participant 3 stated that when reads anything and finds that there are four to five new words, he usually finds the area of the topic and puts a plan to memorize eight to ten words every day in that area. Also, at night when he goes to bed, he tries to practice these words and he found that this strategy was helpful to improve his reading skills. Also, participant 3 stated that he made a plan for two months during which he used to evaluate himself regularly. Also, participant 1 and participant 3 stated that they used to record themselves in order to evaluate their progress in speaking skills through comparing old recordings with the latest recordings. In this regard, participant 1 stated

Mm I mean if you compare the major skills in English like writing, reading, speaking, and listening okay there are aaa specific skills where you can evaluate yourself and you can improve easily by your own. For example, listening and reading is easily evaluated even by computer test, but something like specially speaking, sometime I try to record myself every day and then how aaa I monitor myself progress is by evaluating the last record comparing to the first record. Is there any improvement? This is how I evaluate myself, but I believe in some areas you still need, I mean, you still need to have some feedback from experts.

Another theme is that when self-learners come across difficulties, they always find solutions through searching in the internet. Sometimes they might seek help from experts or other students; however, they often depend on themselves to overcome such difficulties through searching in the internet. In this regard, participant 3 stated

Actually, asking experts in English or searching for in YouTube in channels in aaa other websites that give me the information about this difficulty.

Themes Related to CALL

This section discusses the themes related to CALL elements that are most effective in informal CALL. According to the participants, the first theme is that CALL tools should be suitable for the learners in terms of their level because some websites and language programs are not suitable for the learner's level. The participants argued that not all the websites are suitable for all the learners, as there are websites designed for beginners and websites are suitable for intermediate or advanced students. Participant 1 argued that some websites give a test to help the learner decide if the website is suitable for him; however, when he used such CALL tools, he found that they are not suitable for his level. Also, participant 4 stated that one of the most important elements of websites is stating the targeted students, and whether the website is designed for beginners, intermediate or advanced students. Participant 4 stated

In any website, the most important part is to be suitable for the student or the learner brain or understanding capability. For example, when aaa when you go to the website, it has to be for learners, it has to start with aaa a paragraph or a part of that to convince the student that okay here where you belong to, here where you have to start at.

Also, participant 1 added that the websites usually evaluate the learners' four skills and they give them the same level for the four skills. He stated that sometimes the learner might be in level one in writing, but he is in level three in listening. Therefore, there is a need to guide learners if the CALL tool is suitable for their language level in general and in every skill separately. Besides, participant 4 argued that most of the websites do not provide enough explanation to help the learner decide if the website is suitable for him. He stated that he used to open and practice in the website in order to figure out if it is suitable for him or not.

Another theme is that behaviourist CALL is suitable for English language learners since practice is effective in improving language skills. However, it will be better if the CALL tools start with tutorials to give the learner a background about the skill, followed by exercises and assessment. From the perspective of the participants, the integration of theoretical knowledge and practice helps learners to perform better and improve their English language skills more effectively. In this regard, participant 1 stated

The problem with the websites that are available now now is that some of them will focus totally on exercises, so they they will make the the the I mean the self-study a little bit boring because you have to do exercises all the way. And, sometime the other website will provide you some aaa I mean like aaa self-study without any exercises.

So, this will be missing the exercise while that will be over way with exercises. There is no balance and there is no exchanging in the method of learning.

Another theme is that the participants found CALL tools that use cognitive CALL features effective; however, the materials should not be extremely difficult for the students. The participants argued that they prefer neither easy nor difficult materials. It is better for the materials in the CALL tools to be of some difficulty for them so that they can improve their English language skills. In this regard, participant 1 stated that the materials should be one step above his level so that he can move up to the materials and improve his language skills accordingly. Participant 2 also stated that it is better for the materials to be challenging with practice so that students can improve their language and move to higher levels. Participant 2 stated

Personally, it's ... in the beginning easy materials to, I mean, to prepare yourself to face, I mean, the difficult materials, but with time at the end I referred the advanced materials to test myself I am in the required level or not.

Another theme is related to constructivist CALL which is that the difficulty of the skills should be gradual. That is, the skills should be introduced to the students starting from the easy moving to the difficult gradually. From the perspective of the learners, this gradual practice makes the process of learning smooth. In this regard, participant 2 stated

I mean, not from the, I mean, the beginning using high or advanced level of exercise. It is helpful for the students to not, but to using the easiest things. I mean, during, when you want to go upstairs, you use the first step in the staircase. You cannot jump, I mean, around three meters up.

Also, participant 4 stated

It is a brilliant idea to start from the easy to the hardest. So, starting with easy moving to intermediate, going to the difficult level. Look okay ... what I suggest is when you move from easy to intermediate, for example, at the end of the easy part just test the learner. If you pass the test, go to the intermediate level. And the same way to the top level. So, starting with easy moving to the hard, but being easy all the time it's something ... you will never learn. Being difficult all the time, you will never face like the real English .. go to the real English.

Another theme is related socio-constructivist CALL features which is that it is better if there are communication channels or forums used in the CALL tools so that learners can discuss what they have studied with the other students. Although some of the participants in this study have not practiced this, but they recommend this communication as they feel that learners can exchange ideas and experiences and help each other. The participants also stated that learners can get feedback from experts if there is a channel of communication as this helps to find solutions for their learning difficulties. The participants stated that the communication between learners and teachers or professionals in the field of language is better. They stated that this is not available in most of the websites; however, some websites that require payment provide this service to learners. Participant 1 has elaborated on this point and stated that one of the problems of self-learning is that there are not channels to communicate with others especially professionals to help learners overcome learning obstacles. He stated that he tried such communication a few times through websites that require payment and found it very useful. He added that even in the future, there might be automated communication systems to help learners. In this regard, participant 1 stated

It [feedback] can be through I mean from professionals okay … like you are aaa if you can try to contact them through webeners or aaa live aaa streaming or something like that, or it can be aaa through group discussion with other groups that they are interested in the same topic with you, or it can also be through an automated system if that can be developed within the aaa the aaa the site itself.

To sum up, there are three main themes related to learners. First, the participants have stated that their experience as CALL self-directed learners was very productive and helpful for them to improve their English language. Second, learners need only the basic knowledge for using computers to study English language through CALL. Third, CALL self-directed learners have different strategies to learn English language, and these strategies might be different but usually their experience of studying at the university helped them to use suitable strategies to study English language through CALL. Moreover, there are different themes that are related to CALL according to the experience of the participants as CALL self-directed learners. First, behaviourist CALL is very helpful, and it is better for the CALL tools to provide the learners with tutorials, practice and assessment. Second, cognitive CALL elements are also equally important especially providing learners with challenging materials throughout the process of learning from the first exercise to the last one. Third, constructive CALL elements are also of great importance for the learners, especially building the learners' skills from the beginner level moving ahead to higher levels. Another feature of constructivist CALL which the learners recommended is that CALL tools should save records of evaluation and progress of the learners for their reference so that learners will be able to detect their weaknesses and improve their skills accordingly. Another theme is related to socio-constructivist CALL, which is that communication channels or forums help learners to communicate with others to discuss their learning difficulties. Summary of findings is presented in Table 2 below.

DISCUSSION

The findings of the study are divided into themes under two main categories, namely (a) themes related to learners and (b) themes related to CALL elements. In terms of the themes related to learners, the study found that adult self-directed informal learners have the ICT skills to use technology to study English language. They could manage their learning even when it comes to searching in the internet to find suitable materials. Moreover, adult learners need not training in terms of learning strategies as their previous experience guides them to utilize different learning strategies when they study through CALL. That is, the study at the university equips the students with strategies that can be used to study English language, and most of the learners can use the learning strategies that suit their needs. This finding is against the idea

Table 2. Summary of the findings

Learners' skills needed to formalize informal CALL	
Computer skills	- Learners need basic computer skills to use CALL. - Learners improve their skills to search for CALL materials with time. - There is no need for training to use CALL.
Learning strategies	- Learners use formal learning strategies gained from their study at the university. - Learners use informal learning strategies according to their learning needs.
CALL elements needed to formalize informal CALL	
Behaviourist CALL	- Drill and practice feature is helpful for learners especially when they practice different exercises.
Cognitive CALL	- Challenging materials help learners to improve their language skills. - Attractive features are an important element of CALL.
Constructivist CALL	- Materials that start from the easy skills to difficult skills help learners to improve their language. - Giving tutorials about the skill and then exercises to practice help learners to improve language skills as this links theory to practice. - It is helpful for learners if CALL tools generate reports for learners and save these reports for their reference.
Socio-constructivist CALL	- Communication channels in CALL help learners to find solutions for learning problems. - Communication with teachers through CALL is more helpful for learners. - Communication channels help learners to practice and improve their speaking skills.

of other researchers who stated that learners do have the competencies to turn ICT into a school-like learning (Levinsen & Sørensen, 2008; Gynther, 2010; Sørensen, Audon & Levinsen, 2010 as cited in Levensin & Sørensen, 2011). The experience of the learners proved that they could study independently using CALL to improve their English language skills, and this was helpful for them to reach their goals. This was not achieved randomly as the learners could develop their own strategies to learn and to overcome their learning difficulties. The participants in this study did not depend on teachers or study in a formal setting; however, their learning was self-directed. This leads to the fact that adult self-directed learners have the skills to use technology, and they could develop their learning strategies to make their learning more fruitful. This supports that informal learning and language learning strategies are closely related during the process of learning.

In addition, learning strategies that were employed by the learners varied according to their needs and according to the targeted skills. This makes bricoleur learning strategies an important element for the success of self-directed learning through CALL because learners usually evaluate themselves and use inner dialogue to negotiate and renegotiate their progress to come up with solutions to their learning difficulties. This is in line with the discussion of Levinsen and Sørensen (2011) concerning bricoleur approaches as such approaches link between thinking and practice. This finding is related to learners' skills and learning strategies and it supports the argument of Victori and Lockhart (1995) that adult self-directed learners should be self-directed instead of being instructed. Accordingly, one of the factors for the success of the learners in using CALL to study English language independently is the learners' ICT skills and learning strategies. Hence, in terms of learners' skills to formalize informal CALL, it can be concluded that adult self-directed learners have the required ICT skills and they do not need training. Moreover, in terms of learning strategies, adult self-directed learners can manage their learning as they can evaluate their learning and find solutions to their difficulties. Therefore, learners' ICT skills and learning strategies might not be an issue in formalizing informal CALL among adult learners.

The second category of themes was the elements of CALL that are needed to formalize informal CALL. According to the experience of the learners, all CALL theories are important in the design of CALL. In terms of behaviourist CALL, learners stated that drill and practice is an essential element for the success of their learning. They could polish their language skills because they used to practice different exercises which is the essence of behaviourist CALL (Warchucer & Healey, 1998). However, the learners stated that there should be a balance between theory and practice because learners should understand the skill first and then move to practice which is not available in most of the websites.

In terms of cognitive CALL, the learners also stated that attractive features stimulate their learning and it is important for the materials to be one-step above their level. These features are related to cognitive CALL (Warschauer & Healey, 1998; Hammad & Hussin, 2017; Yassin, Razak & Maasum, 2019) especially that challenging materials helps learners to improve their skills.

In terms of constructivist and socio-constructivist CALL, learners stated that CALL should build the skills of the learners gradually moving from easy skills to difficult skills, and they recommend having communication channels or forums to help learners interact with others as this will be of great help to learners (Warschauer & Healey, 1998; Hammad & Hussin, 2017). The participants in this study generally argued that most of the websites provide practice without theoretical background about the skills. Also, most of the websites do not provide communication channels with others like students, native-speakers and teachers, which is an important aspect of socio-constructivist CALL (Simina & Hamel, 2005). Hence, linking these theories together when designing CALL for English self-directed learners will help to make the process of learning more efficient and productive. When the learners feel that they could get different exercises, link theory to practice, practice in a well-designed website, experience challenging materials that helps them to improve their skills, move gradually from easy to difficult skills, and can communicate with others through to find solutions to their learning difficulties, the process of learning will be more productive and more efficient. This will help to reduce the gap between informal CALL and formal CALL. Such integration will link between theory, technology as well as pedagogy since learners will be directed and supported throughout the process of learning.

IMPLICATIONS

One of the implications of this study is that adult self-directed CALL learners have the skills to use CALL in learning English language skills, and they have acquired learning strategies that make their learning more productive and more fruitful. Therefore, learners are not an issue in formalizing informal CALL.

Another implication is that there is a need to improve CALL tools so that they can suit the needs of the learners. One of the important elements that should be taken into consideration is that CALL tools should focus on the targeted students because this will help learners to find the suitable materials easily. Another thing is that CALL designers should provide feedback according to the obstacles of the learners. That is, different learners might face different difficulties, so CALL tools should provide feedback to learners according to their learning difficulties. Moreover, CALL tools should provide learners with study plan that guides the learners to improve their language. Another important feature is that CALL tools should provide the learners with a record for their performance. This will make the process of learning more effective since the learners will be able to point out their weakness and improve their language skills by focusing in their language weaknesses.

An important implication is that some participants in this study used CALL to study for IELTS, and they could get the desired scores. This provides an implication for universities. That is, in the current age of advanced technology, universities can prepare CALL tools for students. Such CALL tools might have different purposes. For example, universities might use such CALL tools to help students, especially international students, improve their language skills. Besides, universities might make it easier for students to practice and prepare themselves for the admission at the university. Actually, universities might make it easier for students and provide their own tests based on the CALL materials designed by the universities themselves. This will serve the community and students worldwide to improve their language, and students can get admission to these universities easily. This critical because some student might not have the chance to join an institute to study for IELTS or TOEFL because there might not be institutes to teach IELTS or TOEFL in some cities or regions. This idea might globalize education and serve humanity as well.

FUTURE RESEARCH DIRECTIONS

The participants who took part in this study were adult learners doing their higher studies. This explains why adult learners might not need training in ICT skills and learning strategies. Replicating this study with school students might provide different implications concerning ICT skills and learning strategies. Moreover, the nascent 5G internet might make CALL more advanced especially in contexts of communication through CALL. Hence, there is a need for studies on communication channels that address the expectations of the advancement of technology, especially that virtual communication through CALL is going to be a main feature of 5G internet.

CONCLUSION

The difference between informal CALL and formal CALL lies in the setting and the materials. In terms of setting, it cannot be formalized; however, one can formalize informal CALL through a focus on the materials. That is, there is a need to provide learners with CALL tools that take into consideration the elements of behaviourist, cognitive, constructivist and socio-constructivist CALL to make informal CALL formalized, which will maximize the benefit to learners. Moreover, the interaction of the learner with the materials re-directs the focus to be on learners' ICT skills, learning strategies and CALL materials. In terms of learners' ICT skills and learning strategies, they might not be a big deal because using CALL to study English language skills requires the basic computer skills that learners tend to acquire naturally (without formal training). Also, learning strategies might not be an issue with adult self-directed learners because their experience, especially at the university, makes them able to manage their learning using different strategies. However, in terms of CALL tools, there is a need to focus on the features of behaviourist CALL, cognitive CALL, constructivist CALL and socio-constructivist in order to formalize informal CALL. This will also make the process of learning more productive and efficient, and such websites and other CALL tools may reduce the role of the teacher to some extent.

REFERENCES

Astalin, P. K. (2013). Qualitative research designs: A conceptual framework. *International Journal of Social Science and Interdisciplinary Research*, *2*(1), 118–124.

Bahrani, T., & Sim, T. S. (2012). Informal Language Learning Setting: Technology or Social Interaction? *Turkish Online Journal of Educational Technology-TOJET*, *11*(2), 142–149. doi:10.7456/10203100/002

Bahrani, T., Sim, T. S., & Nekoueizadeh, M. (2014). Second language acquisition in informal setting. *Theory and Practice in Language Studies*, *4*(8), 1714–1724. doi:10.4304/tpls.4.8.1714-1723

Beatty, K. (2003). *Teaching and researching computer assisted language learning*. New York: Longman.

Coombs, P. H., & Ahmed, M. (1974). *Attacking Rural Poverty: How Nonformal Education Can Help*. A Research Report for the World Bank Prepared by the International Council for Educational Development.

Crichton, S., & Childs, E. (2005). Clipping and coding audio files: A research method to enable participant voice. *International Journal of Qualitative Methods*, *4*(3), 40–49. doi:10.1177/160940690500400303

Eraut, M. (2004). Informal learning in the workplace. *Studies in Continuing Education*, *26*(2), 247–273. doi:10.1080/158037042000225245

Ferguson, R., Faulkner, D., Whitelock, D., & Sheehy, K. (2015). Pre-teens' informal learning with ICT and Web 2.0. *Technology, Pedagogy and Education*, *24*(2), 247–265. doi:10.1080/1475939X.2013.870596

Garrett, N. (2009). Computer-assisted language learning trends and issues revisited: Integrating innovation. *Modern Language Journal, 93*(s1), 719–740. doi:10.1111/j.1540-4781.2009.00969.x

Gündüz, N. (2005). Computer Assisted Language Learning (CALL). *Journal of Language and Linguistic Studies, 1*(2), 193–214.

Hague, C., & Logan, A. (2009). A review of the current landscape of adult informal learning using digital technologies. *Educational Research.*

Halcomb, E. J., & Davidson, P. M. (2006). Is verbatim transcription of interview data always necessary? *Applied Nursing Research, 19*(1), 38–42. doi:10.1016/j.apnr.2005.06.001 PMID:16455440

Hall, E., Chai, W., & Albrecht, J. A. (2016). A qualitative phenomenological exploration of teachers' experience with nutrition education. *American Journal of Health Education, 47*(3), 136–148. doi:10.1080/19325037.2016.1157532 PMID:27226814

Hammad, Z. M., & Hussin, S. (2017). The Implementation of Computer-Assisted Language Learning in EFL Classroom: Arab EFL Learners Attitudes and Perceptions. *International Journal on Studies in English Language and Literature, 5*(9), 27–37. doi:10.20431/2347-3134.0509004

Intarat, S., Chanchalor, S., & Murphy, E. (2017). ICTs for non-formal education in rural Thailand. *Australasian Journal of Educational Technology, 33*(4), 107–121. doi:10.14742/ajet.3165

Knowles, M. S., Holton, E., & Swanson, R. (2005). *The adult learner: the definitive classic in adult education and human resource development (6th ed.).* Burlington, MA: Elsevier; doi:10.4324/9780080481913

Levinsen, K. T. (2007). Watch out-the Power Users are Coming. *Electronic Journal of e-Learning, 5*(1), 79-86.

Levinsen, K. T., & Sørensen, B. H. (2011). Formalized informal learning: ICT and learning for the 21st century. *International Journal of Digital Literacy and Digital Competence, 2*(1), 7–2. doi:10.4018/jdldc.2011010102

Levy, M. (1997). *Computer-assisted language learning: Context and conceptualization.* Oxford, UK: Oxford University Press.

Lightbown, P. M., & Spada, N. (2001). *How Languages are Learned* (2nd ed.). Oxford, UK: Oxford University Press.

Livingstone, D. W. (2001). *Adults' Informal Learning: Definitions, Findings, Gaps, and Future Research*. NALL Working Paper# 21.

Marsick, V. J., & Watkins, K. (1990). *Informal and Incidental Learning in the Workplace*. London: Routledge.

Merriam, S. B., Caffarella, R. S., & Baumgartner, L. M. (2007). *Learning in adulthood: A comprehensive guide*. John Wiley & Sons.

Moustakas, C. (1994). Phenomenological research methods. *Sage (Atlanta, Ga.)*. doi:10.4135/9781412995658

Nawi, A., Yusoff, A. F. M., Ajmain, H. H., & Abbas, M. R. G. (2014). Engaging student through ICT: Strategies and challenges for using website in teaching and learning. *International Journal on New Trends in Education and Their Implications*, *91*. doi:10.18052/www.scipress.com/ILSHS.27.49

Neal, J. W., Neal, Z. P., VanDyke, E., & Kornbluh, M. (2015). Expediting the analysis of qualitative data in evaluation: A procedure for the rapid identification of themes from audio recordings (RITA). *The American Journal of Evaluation*, *36*(1), 118–132. doi:10.1177/1098214014536601

Okonkwo, U. C. (2011). Computer assisted language learning (CALL) software: Evaluation of its influence in a language learning process. *UJAH: Unizik Journal of Arts and Humanities*, *12*(1), 76–89. doi:10.4314/ujah.v12i1.4

Paredes, R. G., Ogata, H., Saito, N. A., Yin, C., Yno, Y., Oishi, Y., & Ueda, T. (2005, November). LOCH: supporting informal language learning outside the classroom with handhelds. In *IEEE International Workshop on Wireless and Mobile Technologies in Education (WMTE'05)*. IEEE. 10.1109/WMTE.2005.45

Pemberton, L., Fallahkhair, S., & Masthoff, G. (2004). Toward a theoretical framework for informal language learning via interactive television. *Proceeding of CELDA Journal.*

Schwier, R. A. (2010). Focusing educational technology research on informal learning environments. *Contemporary Educational Technology*, *1*(1), 90–92.

Selwyn, N., & Gorard, S. (2004). Exploring the role of ICT in facilitating adult informal learning. *Education Communication and Information*, *4*(2-3), 293–310. doi:10.1080/14636310412331304726

Simina, V., & Hamel, M. J. (2005). CASLA through a social constructivist perspective: WebQuest in project-driven language learning. *ReCALL*, *17*(2), 217–228. doi:10.1017/S0958344005000522

Singh, M. (2009). Recognition, validation and accreditation of non-formal and informal learning and experience: Results of an international study. In *International handbook of education for the changing world of work* (pp. 2597–2613). Dordrecht: Springer. doi:10.1007/978-1-4020-5281-1_171

Tafoli, D., & Golshan, N. (2014). Review of computer-assisted language learning: History, merits and barriers. *International Journal of Language and Linguistics, 2*(5-1), 32-38.

Tessier, S. (2012). From field notes, to transcripts, to tape recordings: Evolution or combination? *International Journal of Qualitative Methods, 11*(4), 446–460. doi:10.1177/160940691201100410

Trinder, R. (2017). Informal and deliberate learning with new technologies. *ELT Journal, 71*(4), 401–412. doi:10.1093/elt/ccw117

Victori, M., & Lockhart, W. (1995). Enhancing metacognition in self-directed language learning. *System, 23*(2), 223–234. doi:10.1016/0346-251X(95)00010-H

Warschauer, M., & Healey, D. (1998). Computers and language learning: An overview. *Language Teaching, 31*(2), 57–71. doi:10.1017/S0261444800012970

Yang, Y. (2010). Computer-assisted foreign language teaching: Theory and Practice. *Journal of Language Teaching and Research, 1*(6), 909–912. doi:10.4304/jltr.1.6.909-912

Yassin, A. A., Razak, N. A., & Maasum, N. R. M. (2019). Investigating the need for computer assisted cooperative learning to improve reading skills among Yemeni university EFL students: A needs analysis study. *International Journal of Virtual and Personal Learning Environments, 9*(2), 15–31. doi:10.4018/IJVPLE.2019070102

Related Readings

To continue IGI Global's long-standing tradition of advancing innovation through emerging research, please find below a compiled list of recommended IGI Global book chapters and journal articles in the areas of language learning, curriculum design, and language interpretation. These related readings will provide additional information and guidance to further enrich your knowledge and assist you with your own research.

Abbadi, S. (2016). Globalization and Possibilities for Intercultural Awareness: Multimodal Arabic Culture Portfolios at a Catholic University. In C. Wang & L. Winstead (Eds.), *Handbook of Research on Foreign Language Education in the Digital Age* (pp. 277–294). Hershey, PA: IGI Global. doi:10.4018/978-1-5225-0177-0.ch013

Ahluwalia, G., & Gupta, D. (2017). Impact of Technology-Enhanced Language Learning on the Writing Skills of Engineering Students: A Case Study. In D. Tafazoli & M. Romero (Eds.), *Multiculturalism and Technology-Enhanced Language Learning* (pp. 33–56). Hershey, PA: IGI Global. doi:10.4018/978-1-5225-1882-2.ch003

Akayoglu, S., & Seferoglu, G. (2017). Social Presence Functions in Task-Based Language Activities in a Virtual Classroom in Second Life. In D. Tafazoli & M. Romero (Eds.), *Multiculturalism and Technology-Enhanced Language Learning* (pp. 181–198). Hershey, PA: IGI Global. doi:10.4018/978-1-5225-1882-2.ch011

Andrawiss-Dlamini, H. M., & Puntil, D. (2018). Open Up the Classroom: The Use of Film Excerpts and Video Clips in Lower Levels Language Courses. In C. Xiang (Ed.), *Cases on Audio-Visual Media in Language Education* (pp. 22–46). Hershey, PA: IGI Global. doi:10.4018/978-1-5225-2724-4.ch002

Anwaruddin, S. M. (2018). Beyond Determinism and Instrumentalism: Re-Conceptualizing Technology for CALL. In J. Perren, K. Kelch, J. Byun, S. Cervantes, & S. Safavi (Eds.), *Applications of CALL Theory in ESL and EFL Environments* (pp. 22–35). Hershey, PA: IGI Global. doi:10.4018/978-1-5225-2933-0.ch002

Ashtari, N. (2018). Mixed Reality Environments in Teaching and Learning English. In D. Tafazoli, M. Gomez Parra, & C. Huertas-Abril (Eds.), *Cross-Cultural Perspectives on Technology-Enhanced Language Learning* (pp. 187–202). Hershey, PA: IGI Global. doi:10.4018/978-1-5225-5463-9.ch011

Awada, G. M., Diab, H. B., & Faour, K. H. (2018). Effect of GI and Glogster on Improving the Intercultural Communication Skills in Higher Education. In D. Tafazoli, M. Gomez Parra, & C. Huertas-Abril (Eds.), *Cross-Cultural Perspectives on Technology-Enhanced Language Learning* (pp. 130–158). Hershey, PA: IGI Global. doi:10.4018/978-1-5225-5463-9.ch008

Azad, M. (2017). Dynamic Assessment and Multi-Media Language Instruction. In D. Tafazoli & M. Romero (Eds.), *Multiculturalism and Technology-Enhanced Language Learning* (pp. 230–251). Hershey, PA: IGI Global. doi:10.4018/978-1-5225-1882-2.ch014

Balchin, K., & Wild, C. (2018). It's All in the Numbers: Enhancing Technology Use in Urban and Rural Environments. In D. Tafazoli, M. Gomez Parra, & C. Huertas-Abril (Eds.), *Cross-Cultural Perspectives on Technology-Enhanced Language Learning* (pp. 203–221). Hershey, PA: IGI Global. doi:10.4018/978-1-5225-5463-9.ch012

Bañares-Marivela, E., & Rayón-Rumayor, L. (2017). Multimodal Narratives and iPad in Second Language Teaching. In D. Tafazoli & M. Romero (Eds.), *Multiculturalism and Technology-Enhanced Language Learning* (pp. 57–79). Hershey, PA: IGI Global. doi:10.4018/978-1-5225-1882-2.ch004

Barnes, M. (2017). Encouraging Communication through the Use of Educational Social Media Tools. In D. Tafazoli & M. Romero (Eds.), *Multiculturalism and Technology-Enhanced Language Learning* (pp. 1–12). Hershey, PA: IGI Global. doi:10.4018/978-1-5225-1882-2.ch001

Barrons, K. P. (2017). Flipped Instructional Technology: Developing MIS Competencies Applying Enterprise Resource Planning Software. In J. Loucky & J. Ware (Eds.), *Flipped Instruction Methods and Digital Technologies in the Language Learning Classroom* (pp. 183–195). Hershey, PA: IGI Global. doi:10.4018/978-1-5225-0824-3.ch008

Bekleyen, N., & Çelik, S. (2017). Attitudes of Adult EFL Learners towards Preparing for a Language Test via CALL. In D. Tafazoli & M. Romero (Eds.), *Multiculturalism and Technology-Enhanced Language Learning* (pp. 214–229). Hershey, PA: IGI Global. doi:10.4018/978-1-5225-1882-2.ch013

Blattner, G., Dalola, A., & Lomicka, L. (2016). Mind Your Hashtags: A Sociopragmatic Study of Student Interpretations of French Native Speakers' Tweets. In C. Wang & L. Winstead (Eds.), *Handbook of Research on Foreign Language Education in the Digital Age* (pp. 33–58). Hershey, PA: IGI Global. doi:10.4018/978-1-5225-0177-0.ch003

Bobkina, J., & Romero, E. D. (2018). Multimodal Listening Through Movie Trailers: Towards a Framework for Classroom Implementation. In C. Xiang (Ed.), *Cases on Audio-Visual Media in Language Education* (pp. 69–92). Hershey, PA: IGI Global. doi:10.4018/978-1-5225-2724-4.ch004

Bodah, E. T., Alldredge, J. M., Bodah, B. W., Neckel, A., & Goellner, E. (2016). Challenges and Perspectives of Language Education Technology in Brazil: From Confronting Native Language Loss to Implementing EFL Classes. In C. Wang & L. Winstead (Eds.), *Handbook of Research on Foreign Language Education in the Digital Age* (pp. 59–82). Hershey, PA: IGI Global. doi:10.4018/978-1-5225-0177-0.ch004

Bourgoin, R. C. (2016). French Immersion "So Why Would You do Something Like That to a Child?": Issues of Advocacy, Accessibility, and Inclusion. *International Journal of Bias, Identity and Diversities in Education, 1*(1), 42–58. doi:10.4018/IJBIDE.2016010104

Brown, D. (2016). Motivation and Learning Outcomes in Reciprocal SCMC Language-Learning. *International Journal of Computer-Assisted Language Learning and Teaching, 6*(2), 19–34. doi:10.4018/IJCALLT.2016040102

Burgo, C. (2016). Teaching Spanish in the Digital Age: A Flipped Classroom or Just Hybrid? In C. Wang & L. Winstead (Eds.), *Handbook of Research on Foreign Language Education in the Digital Age* (pp. 83–103). Hershey, PA: IGI Global. doi:10.4018/978-1-5225-0177-0.ch005

Carter, B., He, M., & Kawasaki, T. (2018). Using YouTube and WeChat to Promote Communicative and Intercultural Competence in Japanese and Mandarin. In C. Xiang (Ed.), *Cases on Audio-Visual Media in Language Education* (pp. 219–239). Hershey, PA: IGI Global. doi:10.4018/978-1-5225-2724-4.ch009

Cho, H. (2018). The Pitfalls and Promises of Electronic Portfolio Assessment With Secondary English Language Learners. In J. Perren, K. Kelch, J. Byun, S. Cervantes, & S. Safavi (Eds.), *Applications of CALL Theory in ESL and EFL Environments* (pp. 111–130). Hershey, PA: IGI Global. doi:10.4018/978-1-5225-2933-0.ch007

Cimermanova, I. (2017). Computer-Assisted Learning in Slovakia. In D. Tafazoli & M. Romero (Eds.), *Multiculturalism and Technology-Enhanced Language Learning* (pp. 252–270). Hershey, PA: IGI Global. doi:10.4018/978-1-5225-1882-2.ch015

Dalton, G., & Devitt, A. (2016). Gaeilge Gaming: Assessing how games can help children to learn Irish. *International Journal of Game-Based Learning*, *6*(4), 22–38. doi:10.4018/IJGBL.2016100102

Dinh, H. (2018). Teachers' Uses and Beliefs in the Integration of YouTube Videos Into English Language Teaching: A Comparison Between ESL and EFL Contexts. In J. Perren, K. Kelch, J. Byun, S. Cervantes, & S. Safavi (Eds.), *Applications of CALL Theory in ESL and EFL Environments* (pp. 94–110). Hershey, PA: IGI Global. doi:10.4018/978-1-5225-2933-0.ch006

Dixon, D. H., & Christison, M. (2018). The Usefulness of Massive Multiplayer Online Role Playing Games (MMORPGs) as Tools for Promoting Second Language Acquisition. In J. Perren, K. Kelch, J. Byun, S. Cervantes, & S. Safavi (Eds.), *Applications of CALL Theory in ESL and EFL Environments* (pp. 244–268). Hershey, PA: IGI Global. doi:10.4018/978-1-5225-2933-0.ch014

Dixon, T., & Christison, M. (2018). Teaching English Grammar in a Hybrid Academic ESL Course. In J. Perren, K. Kelch, J. Byun, S. Cervantes, & S. Safavi (Eds.), *Applications of CALL Theory in ESL and EFL Environments* (pp. 149–169). Hershey, PA: IGI Global. doi:10.4018/978-1-5225-2933-0.ch009

du Plessis, J. (2016). Reflection: How Now Shapes the Future – Emerging Trends from the LCTL Trenches. In C. Wang & L. Winstead (Eds.), *Handbook of Research on Foreign Language Education in the Digital Age* (pp. 1–8). Hershey, PA: IGI Global. doi:10.4018/978-1-5225-0177-0.ch001

Durand, S. (2018). Embracing Computer Corpora in the Language Learning Classroom and Using It in Your Classroom. In J. Perren, K. Kelch, J. Byun, S. Cervantes, & S. Safavi (Eds.), *Applications of CALL Theory in ESL and EFL Environments* (pp. 131–148). Hershey, PA: IGI Global. doi:10.4018/978-1-5225-2933-0.ch008

Foley-Ryan, M. M. (2018). Using News Reports in Language Education. In C. Xiang (Ed.), *Cases on Audio-Visual Media in Language Education* (pp. 47–68). Hershey, PA: IGI Global. doi:10.4018/978-1-5225-2724-4.ch003

Forsythe, E. (2017). Integrating Recent CALL Innovations into Flipped Instruction. In J. Loucky & J. Ware (Eds.), *Flipped Instruction Methods and Digital Technologies in the Language Learning Classroom* (pp. 270–277). Hershey, PA: IGI Global. doi:10.4018/978-1-5225-0824-3.ch012

Forsythe, E. (2017). Pedagogical Rationale for Flipped Learning and Digital Technology in Second Language Acquisition. In J. Loucky & J. Ware (Eds.), *Flipped Instruction Methods and Digital Technologies in the Language Learning Classroom* (pp. 1–20). Hershey, PA: IGI Global. doi:10.4018/978-1-5225-0824-3.ch001

Fuchs, C. (2016). The Structural and Dialogic Aspects of Language Massive Open Online Courses (LMOOCs): A Case Study. In C. Wang & L. Winstead (Eds.), *Handbook of Research on Foreign Language Education in the Digital Age* (pp. 10–32). Hershey, PA: IGI Global. doi:10.4018/978-1-5225-0177-0.ch002

Gabaudan, O. R. (2017). Students as Producers of Digital Audio-Visual Clips in Preparation for a Study Abroad Period: Collaboration for Independent Language Learning. In D. Tafazoli & M. Romero (Eds.), *Multiculturalism and Technology-Enhanced Language Learning* (pp. 97–115). Hershey, PA: IGI Global. doi:10.4018/978-1-5225-1882-2.ch006

Galvez, H. (2017). Flipped Instruction for Language Learning. In J. Loucky & J. Ware (Eds.), *Flipped Instruction Methods and Digital Technologies in the Language Learning Classroom* (pp. 91–107). Hershey, PA: IGI Global. doi:10.4018/978-1-5225-0824-3.ch004

Garnes-Tarazona, I. (2018). Exploring Socio-Cultural Elements in Three Commercial English Language Learning Apps. *International Journal of Computer-Assisted Language Learning and Teaching*, *8*(1), 65–82. doi:10.4018/IJCALLT.2018010105

Gasmi, A. A., & Thomas, M. (2017). Academic Writing in the Flipped EFL Classroom: A Case Study on Student Engagement in Oman. In J. Loucky & J. Ware (Eds.), *Flipped Instruction Methods and Digital Technologies in the Language Learning Classroom* (pp. 232–251). Hershey, PA: IGI Global. doi:10.4018/978-1-5225-0824-3.ch010

Gremler, C., & Wielander, E. (2018). The Benefits of Student-Led Video Production in the Language for Business Classroom. In C. Xiang (Ed.), *Cases on Audio-Visual Media in Language Education* (pp. 155–193). Hershey, PA: IGI Global. doi:10.4018/978-1-5225-2724-4.ch007

Heirati, J. K., Azadsarv, M., & Golzari, A. (2017). Technology-Enhanced Vocabulary Development through Semantic Network Development Model vs. Syntactic Structure Development Model. In D. Tafazoli & M. Romero (Eds.), *Multiculturalism and Technology-Enhanced Language Learning* (pp. 199–213). Hershey, PA: IGI Global. doi:10.4018/978-1-5225-1882-2.ch012

Ibrahim, K. H. (2018). The Foreign Language Learning Potential of Video Games: FL Games as Cross-Cultural Texts, Narratives, and Artifacts. In D. Tafazoli, M. Gomez Parra, & C. Huertas-Abril (Eds.), *Cross-Cultural Perspectives on Technology-Enhanced Language Learning* (pp. 159–171). Hershey, PA: IGI Global. doi:10.4018/978-1-5225-5463-9.ch009

Johnston, K. P., & Lawrence, G. (2018). A Theoretically Informed Approach to Collaborative Writing in EAP Contexts Using Web-Based Technologies. In J. Perren, K. Kelch, J. Byun, S. Cervantes, & S. Safavi (Eds.), *Applications of CALL Theory in ESL and EFL Environments* (pp. 1–21). Hershey, PA: IGI Global. doi:10.4018/978-1-5225-2933-0.ch001

Kan, Q., & Tang, J. (2018). Researching Mobile-Assisted English Language Learning Among Adult Distance Learners in China: Emerging Practices and Learner Perception of Teacher Role. *International Journal of Computer-Assisted Language Learning and Teaching*, *8*(3), 1–28. doi:10.4018/IJCALLT.2018070101

Kasemsap, K. (2017). Mastering Technology-Enhanced Language Learning, Computer-Assisted Language Learning, and Mobile-Assisted Language Learning. In D. Tafazoli & M. Romero (Eds.), *Multiculturalism and Technology-Enhanced Language Learning* (pp. 157–180). Hershey, PA: IGI Global. doi:10.4018/978-1-5225-1882-2.ch010

Katrnakova, H. (2018). Integration of Videoconferencing in ESP and EAP Courses: Focus on Community of Practice Interaction. In C. Xiang (Ed.), *Cases on Audio-Visual Media in Language Education* (pp. 283–303). Hershey, PA: IGI Global. doi:10.4018/978-1-5225-2724-4.ch012

Kirk, S., & Casenove, D. (2017). Flipping the Academic Writing Classroom. In J. Loucky & J. Ware (Eds.), *Flipped Instruction Methods and Digital Technologies in the Language Learning Classroom* (pp. 196–231). Hershey, PA: IGI Global. doi:10.4018/978-1-5225-0824-3.ch009

Kızıl, A. Ş., & Kilimci, A. (2017). Teaching Collocations through Web-Based Concordancing: A Corpus Informed Quasi-Experimental Study. In D. Tafazoli & M. Romero (Eds.), *Multiculturalism and Technology-Enhanced Language Learning* (pp. 142–156). Hershey, PA: IGI Global. doi:10.4018/978-1-5225-1882-2.ch009

Kostka, I., & Marshall, H. W. (2018). Flipped Learning in TESOL: Past, Present, and Future. In J. Perren, K. Kelch, J. Byun, S. Cervantes, & S. Safavi (Eds.), *Applications of CALL Theory in ESL and EFL Environments* (pp. 223–243). Hershey, PA: IGI Global. doi:10.4018/978-1-5225-2933-0.ch013

Lawrence, G., & Ahmed, F. (2018). Pedagogical Insights into Hyper-Immersive Virtual World Language Learning Environments. *International Journal of Computer-Assisted Language Learning and Teaching, 8*(1), 1–14. doi:10.4018/IJCALLT.2018010101

Lawrence, G., & Spector-Cohen, E. (2018). Examining International Telecollaboration in Language Teacher Education. In D. Tafazoli, M. Gomez Parra, & C. Huertas-Abril (Eds.), *Cross-Cultural Perspectives on Technology-Enhanced Language Learning* (pp. 322–345). Hershey, PA: IGI Global. doi:10.4018/978-1-5225-5463-9.ch018

Le, N., & Dinh, H. (2018). Augmented Reality: A Brief Introduction, Its Potentials, and Implications in Language Education. In J. Perren, K. Kelch, J. Byun, S. Cervantes, & S. Safavi (Eds.), *Applications of CALL Theory in ESL and EFL Environments* (pp. 291–309). Hershey, PA: IGI Global. doi:10.4018/978-1-5225-2933-0.ch016

Lee, L. (2018). Using Telecollaboration 2.0 to Build Intercultural Communicative Competence: A Spanish-American Exchange. In D. Tafazoli, M. Gomez Parra, & C. Huertas-Abril (Eds.), *Cross-Cultural Perspectives on Technology-Enhanced Language Learning* (pp. 303–321). Hershey, PA: IGI Global. doi:10.4018/978-1-5225-5463-9.ch017

Legutko, A. (2016). Yiddish in the 21st Century: New Media to the Rescue of Endangered Languages. In C. Wang & L. Winstead (Eds.), *Handbook of Research on Foreign Language Education in the Digital Age* (pp. 236–252). Hershey, PA: IGI Global. doi:10.4018/978-1-5225-0177-0.ch011

Levy, T., & Cohen, L. J. (2017). Teaching English for Academic Purposes through the Use of Digital Tools. In J. Loucky & J. Ware (Eds.), *Flipped Instruction Methods and Digital Technologies in the Language Learning Classroom* (pp. 21–63). Hershey, PA: IGI Global. doi:10.4018/978-1-5225-0824-3.ch002

Lewis, T. N. (2016). Creating a Micro-Immersion Environment Through Telecollaboration. In C. Wang & L. Winstead (Eds.), *Handbook of Research on Foreign Language Education in the Digital Age* (pp. 144–169). Hershey, PA: IGI Global. doi:10.4018/978-1-5225-0177-0.ch007

Li, B., Yao, S., & Hong, W. (2016). Beginning Chinese as a Foreign Language Online Course Design: Utilizing Multiple Digital Modes and Assessments. In C. Wang & L. Winstead (Eds.), *Handbook of Research on Foreign Language Education in the Digital Age* (pp. 104–142). Hershey, PA: IGI Global. doi:10.4018/978-1-5225-0177-0.ch006

Lim, B., & Pyun, D. O. (2016). Korean Foreign Language Learning: Videoconferencing with Native Speakers. In C. Wang & L. Winstead (Eds.), *Handbook of Research on Foreign Language Education in the Digital Age* (pp. 253–276). Hershey, PA: IGI Global. doi:10.4018/978-1-5225-0177-0.ch012

Liontas, J. I. (2018). Understanding Idiomaticity in CALL. In J. Perren, K. Kelch, J. Byun, S. Cervantes, & S. Safavi (Eds.), *Applications of CALL Theory in ESL and EFL Environments* (pp. 36–58). Hershey, PA: IGI Global. doi:10.4018/978-1-5225-2933-0.ch003

Liu, H., Tao, W., & Cain, W. (2016). Investigating Mobile Assisted English Foreign Language Learning and Teaching in China: Issues, Attitudes and Perceptions. In C. Wang & L. Winstead (Eds.), *Handbook of Research on Foreign Language Education in the Digital Age* (pp. 315–333). Hershey, PA: IGI Global. doi:10.4018/978-1-5225-0177-0.ch015

Llull, J., Palacios, A., & Ulkuniemi, S. (2018). Fostering Intercultural Competence Through Art and ICT Among University Students in Spain and Finland. In D. Tafazoli, M. Gomez Parra, & C. Huertas-Abril (Eds.), *Cross-Cultural Perspectives on Technology-Enhanced Language Learning* (pp. 237–262). Hershey, PA: IGI Global. doi:10.4018/978-1-5225-5463-9.ch014

Loucky, J. P. (2017). Motivating and Empowering Students' Language Learning in Flipped Integrated English Classes. In J. Loucky & J. Ware (Eds.), *Flipped Instruction Methods and Digital Technologies in the Language Learning Classroom* (pp. 108–138). Hershey, PA: IGI Global. doi:10.4018/978-1-5225-0824-3.ch005

Loucky, J. P. (2017). Studies of Flipping Classes with Asian Students. In J. Loucky & J. Ware (Eds.), *Flipped Instruction Methods and Digital Technologies in the Language Learning Classroom* (pp. 64–90). Hershey, PA: IGI Global. doi:10.4018/978-1-5225-0824-3.ch003

Loucky, J. P. (2017). Testing a CALL Effectiveness Model: Online Media Can Open New Learning Possibilities. In J. Loucky & J. Ware (Eds.), *Flipped Instruction Methods and Digital Technologies in the Language Learning Classroom* (pp. 252–269). Hershey, PA: IGI Global. doi:10.4018/978-1-5225-0824-3.ch011

Luan, D. (2018). Exploring Intercultural Communication Through the Act of Translation in Subtitling. In C. Xiang (Ed.), *Cases on Audio-Visual Media in Language Education* (pp. 261–281). Hershey, PA: IGI Global. doi:10.4018/978-1-5225-2724-4.ch011

Martínez-Carrasco, R. (2018). Collaborative Writing 2.0: Socializing Critical, Cross-Cultural Agents Through Online, Project-Based Methodology. In D. Tafazoli, M. Gomez Parra, & C. Huertas-Abril (Eds.), *Cross-Cultural Perspectives on Technology-Enhanced Language Learning* (pp. 283–302). Hershey, PA: IGI Global. doi:10.4018/978-1-5225-5463-9.ch016

Mary, L., & Young, A. (2016). The Role of Multi-Media in Expanding Pre-Service Teachers' Understanding of Culturally and Linguistically Diverse Classrooms and Furthering Their Professional Identities. In C. Wang & L. Winstead (Eds.), *Handbook of Research on Foreign Language Education in the Digital Age* (pp. 296–314). Hershey, PA: IGI Global. doi:10.4018/978-1-5225-0177-0.ch014

Melo-Pfeifer, S. (2016). Translanguaging in Multilingual Chat Interaction: Opportunities for Intercomprehension between Romance Languages. In C. Wang & L. Winstead (Eds.), *Handbook of Research on Foreign Language Education in the Digital Age* (pp. 188–207). Hershey, PA: IGI Global. doi:10.4018/978-1-5225-0177-0.ch009

Mihăeş, L. C. (2018). Technology-Integrated Teaching and Assessment: ESP at the University of Bucharest. In J. Perren, K. Kelch, J. Byun, S. Cervantes, & S. Safavi (Eds.), *Applications of CALL Theory in ESL and EFL Environments* (pp. 187–205). Hershey, PA: IGI Global. doi:10.4018/978-1-5225-2933-0.ch011

Mohammadi, E., & Shirkamar, Z. S. (2018). Mobile-Assisted Language Learning: Challenges and Setbacks in Developing Countries. In D. Tafazoli, M. Gomez Parra, & C. Huertas-Abril (Eds.), *Cross-Cultural Perspectives on Technology-Enhanced Language Learning* (pp. 172–186). Hershey, PA: IGI Global. doi:10.4018/978-1-5225-5463-9.ch010

Monfared, A., Cervantes, S. E., Lee, S. M., & Jackson, M. (2018). Establishing a Praxis Between SLA Theory and CALL-Based Practices. In J. Perren, K. Kelch, J. Byun, S. Cervantes, & S. Safavi (Eds.), *Applications of CALL Theory in ESL and EFL Environments* (pp. 59–79). Hershey, PA: IGI Global. doi:10.4018/978-1-5225-2933-0.ch004

Morgana, V., & Shrestha, P. N. (2018). Investigating Students' and Teachers' Perceptions of Using the iPad in an Italian English as a Foreign Language Classroom. *International Journal of Computer-Assisted Language Learning and Teaching*, *8*(3), 29–49. doi:10.4018/IJCALLT.2018070102

Murtiningsih, S. R., & Rahmawati, F. (2018). A CALL-Mediated Course to Enhance EFL Pre-Service Teachers' Engagement. In J. Perren, K. Kelch, J. Byun, S. Cervantes, & S. Safavi (Eds.), *Applications of CALL Theory in ESL and EFL Environments* (pp. 170–186). Hershey, PA: IGI Global. doi:10.4018/978-1-5225-2933-0.ch010

Naidoo, V. (2017). Challenges Facing E-Learning. In D. Tafazoli & M. Romero (Eds.), *Multiculturalism and Technology-Enhanced Language Learning* (pp. 271–288). Hershey, PA: IGI Global. doi:10.4018/978-1-5225-1882-2.ch016

Newcombe, J., & Brick, B. (2017). Blending Video Games Into Language Learning. *International Journal of Computer-Assisted Language Learning and Teaching*, *7*(4), 75–89. doi:10.4018/IJCALLT.2017100106

Olesova, L., & de Oliveira, L. (2018). Using Feedback in ESL and EFL Asynchronous Online Environments. In J. Perren, K. Kelch, J. Byun, S. Cervantes, & S. Safavi (Eds.), *Applications of CALL Theory in ESL and EFL Environments* (pp. 206–222). Hershey, PA: IGI Global. doi:10.4018/978-1-5225-2933-0.ch012

Pan, F. (2018). Short Video Production for Language Learning and Assessment. In C. Xiang (Ed.), *Cases on Audio-Visual Media in Language Education* (pp. 194–217). Hershey, PA: IGI Global. doi:10.4018/978-1-5225-2724-4.ch008

Piri, S., & Riahi, S. (2018). Cross-Cultural Perspectives on Technology-Enhanced Language Learning: A Review of Research. In D. Tafazoli, M. Gomez Parra, & C. Huertas-Abril (Eds.), *Cross-Cultural Perspectives on Technology-Enhanced Language Learning* (pp. 1–15). Hershey, PA: IGI Global. doi:10.4018/978-1-5225-5463-9.ch001

Przymus, S. D., & Smith, A. R. (2018). Eres un Gamer?: Engaging Transnational Children in Game-Ecology Language and Identity Socialization Within the EFL Environment. In J. Perren, K. Kelch, J. Byun, S. Cervantes, & S. Safavi (Eds.), *Applications of CALL Theory in ESL and EFL Environments* (pp. 269–290). Hershey, PA: IGI Global. doi:10.4018/978-1-5225-2933-0.ch015

Ramírez-Verdugo, M. D., & García de la Vega, A. (2018). A Conceptual Reference Framework for Sustainability Education in Multilingual and Cross-Cultural Settings: Applied Technology, Transmedia, and Digital Storytelling. In D. Tafazoli, M. Gomez Parra, & C. Huertas-Abril (Eds.), *Cross-Cultural Perspectives on Technology-Enhanced Language Learning* (pp. 222–236). Hershey, PA: IGI Global. doi:10.4018/978-1-5225-5463-9.ch013

Rao, Y., Wang, C., & Bender, J. (2016). French-Chinese Dialogical Interaction via Web Collaborative Blog-Writing: Code-Switching to Extend Online Tandem Language Learning. In C. Wang & L. Winstead (Eds.), *Handbook of Research on Foreign Language Education in the Digital Age* (pp. 208–234). Hershey, PA: IGI Global. doi:10.4018/978-1-5225-0177-0.ch010

Reitz, L., Sohny, A., & Lochmann, G. (2016). VR-Based Gamification of Communication Training and Oral Examination in a Second Language. *International Journal of Game-Based Learning*, *6*(2), 46–61. doi:10.4018/IJGBL.2016040104

Roberto, J. A. (2018). Cross-Cultural Languages Behind Technology-Enhanced Language Learning. In D. Tafazoli, M. Gomez Parra, & C. Huertas-Abril (Eds.), *Cross-Cultural Perspectives on Technology-Enhanced Language Learning* (pp. 16–46). Hershey, PA: IGI Global. doi:10.4018/978-1-5225-5463-9.ch002

Rosen, L. B., Maeda, M., & Roberts, N. (2017). Gain Time and Differentiate to Meet Student Needs in University Learning Environments: A Flipped Learning Approach. In J. Loucky & J. Ware (Eds.), *Flipped Instruction Methods and Digital Technologies in the Language Learning Classroom* (pp. 159–182). Hershey, PA: IGI Global. doi:10.4018/978-1-5225-0824-3.ch007

Sadeghi, A. R., & Ghorbani, S. (2017). The Impact of TED-Vodcast on Iranian EFL Learners' Academic Oral Proficiency. In D. Tafazoli & M. Romero (Eds.), *Multiculturalism and Technology-Enhanced Language Learning* (pp. 80–96). Hershey, PA: IGI Global. doi:10.4018/978-1-5225-1882-2.ch005

Sadoux, M. (2018). Life-Long Language Learning Strategies for a Brave New Digital World: Collaborative Design and Delivery of an Online Module. In C. Xiang (Ed.), *Cases on Audio-Visual Media in Language Education* (pp. 350–377). Hershey, PA: IGI Global. doi:10.4018/978-1-5225-2724-4.ch015

Sahragard, R., & Meihami, H. (2017). A Sociocultural Perspective into Intercultural Competence and Computer-Assisted Language Learning: Intercultural Competence and Computer-Assisted Language Learning. In D. Tafazoli & M. Romero (Eds.), *Multiculturalism and Technology-Enhanced Language Learning* (pp. 130–141). Hershey, PA: IGI Global. doi:10.4018/978-1-5225-1882-2.ch008

Saito, R., Thor, M. H., & Inose, H. (2018). Developing Intercultural Competence and Language Skills Through International Online Collaborative Learning. In C. Xiang (Ed.), *Cases on Audio-Visual Media in Language Education* (pp. 304–327). Hershey, PA: IGI Global. doi:10.4018/978-1-5225-2724-4.ch013

Santamaría-García, C. (2018). Connected Learners: Online and Off-Line Learning With a Focus on Politeness Intercultural Competences. In D. Tafazoli, M. Gomez Parra, & C. Huertas-Abril (Eds.), *Cross-Cultural Perspectives on Technology-Enhanced Language Learning* (pp. 83–99). Hershey, PA: IGI Global. doi:10.4018/978-1-5225-5463-9.ch005

Shahrokhi, M., & Taheri, S. (2016). The Impact of Blog Peer Feedback on Improving Iranian English Foreign Language Students' Writing. In C. Wang & L. Winstead (Eds.), *Handbook of Research on Foreign Language Education in the Digital Age* (pp. 365–386). Hershey, PA: IGI Global. doi:10.4018/978-1-5225-0177-0.ch017

Shiau, H., & Xiang, C. H. (2018). Learning English With Texting on Social Media: The Case of Taiwanese ESL Students in the U.S. In C. Xiang (Ed.), *Cases on Audio-Visual Media in Language Education* (pp. 240–260). Hershey, PA: IGI Global. doi:10.4018/978-1-5225-2724-4.ch010

Shin, D. (2018). Contextual Factors of Online Teacher Education in TESOL: An Activity Theory Perspective. In J. Perren, K. Kelch, J. Byun, S. Cervantes, & S. Safavi (Eds.), *Applications of CALL Theory in ESL and EFL Environments* (pp. 80–93). Hershey, PA: IGI Global. doi:10.4018/978-1-5225-2933-0.ch005

Smith, A. E., & O'Brien, D. (2018). Use of Videos in Teaching and Learning: Creating Language Video Resources for British Sign Language (BSL). In C. Xiang (Ed.), *Cases on Audio-Visual Media in Language Education* (pp. 1–21). Hershey, PA: IGI Global. doi:10.4018/978-1-5225-2724-4.ch001

Solmaz, O. (2018). Social Networking Sites: Cross-Cultural Perspectives, Implications, and Applicable Frameworks for L2 Teaching and Learning. In D. Tafazoli, M. Gomez Parra, & C. Huertas-Abril (Eds.), *Cross-Cultural Perspectives on Technology-Enhanced Language Learning* (pp. 118–129). Hershey, PA: IGI Global. doi:10.4018/978-1-5225-5463-9.ch007

Stannard, R., & Mann, S. (2018). Using Screen Capture Feedback to Establish Social Presence and Increase Student Engagement: A Genuine Innovation in Feedback. In C. Xiang (Ed.), *Cases on Audio-Visual Media in Language Education* (pp. 93–116). Hershey, PA: IGI Global. doi:10.4018/978-1-5225-2724-4.ch005

Stepanek, L., & Arriaga, C. G. (2018). Video Summaries of Academic Texts. In C. Xiang (Ed.), *Cases on Audio-Visual Media in Language Education* (pp. 328–349). Hershey, PA: IGI Global. doi:10.4018/978-1-5225-2724-4.ch014

Teng, F. (2017). A New Era of Applying CALL to Enhance EFL Learners' Lexical Knowledge. In J. Loucky & J. Ware (Eds.), *Flipped Instruction Methods and Digital Technologies in the Language Learning Classroom* (pp. 139–158). Hershey, PA: IGI Global. doi:10.4018/978-1-5225-0824-3.ch006

Vinagre, M. (2016). Developing Key Competences for Life-Long Learning through Virtual Collaboration: Teaching ICT in English as a Medium of Instruction. In C. Wang & L. Winstead (Eds.), *Handbook of Research on Foreign Language Education in the Digital Age* (pp. 170–187). Hershey, PA: IGI Global. doi:10.4018/978-1-5225-0177-0.ch008

Vurdien, R. (2017). Mobile Assisted Vocabulary Acquisition and Wikis to Enhance Writing Skills. *International Journal of Computer-Assisted Language Learning and Teaching*, *7*(2), 1–21. doi:10.4018/IJCALLT.2017040101

Vurdien, R., & Puranen, P. (2016). A Spanish-Finnish Telecollaboration to Develop Intercultural Competence and Learner Autonomy. *International Journal of Computer-Assisted Language Learning and Teaching*, *6*(3), 36–53. doi:10.4018/IJCALLT.2016070103

Vurdien, R., & Puranen, P. (2018). Intercultural Learning via Videoconferencing: Students' Attitudes and Experiences. In D. Tafazoli, M. Gomez Parra, & C. Huertas-Abril (Eds.), *Cross-Cultural Perspectives on Technology-Enhanced Language Learning* (pp. 264–282). Hershey, PA: IGI Global. doi:10.4018/978-1-5225-5463-9.ch015

Wang, A. (2018). Issues of Cross-Cultural Communications in a Globalizing Era. In D. Tafazoli, M. Gomez Parra, & C. Huertas-Abril (Eds.), *Cross-Cultural Perspectives on Technology-Enhanced Language Learning* (pp. 100–116). Hershey, PA: IGI Global. doi:10.4018/978-1-5225-5463-9.ch006

Werner, V., Lehl, M., & Walton, J. (2017). Pop Lyrics and Mobile Language Learning: Prospects and Challenges. *International Journal of Mobile and Blended Learning*, *9*(4), 33–48. doi:10.4018/IJMBL.2017100103

White, J. R. (2018). Marking Community Identity Through Languaging: Authentic Norms in TELL. In D. Tafazoli, M. Gomez Parra, & C. Huertas-Abril (Eds.), *Cross-Cultural Perspectives on Technology-Enhanced Language Learning* (pp. 47–67). Hershey, PA: IGI Global. doi:10.4018/978-1-5225-5463-9.ch003

Winstead, L. (2016). Mexican Heritage ELL and Native English Speaker Interaction: A Case Study of Tandem Language Learning Strategies. In C. Wang & L. Winstead (Eds.), *Handbook of Research on Foreign Language Education in the Digital Age* (pp. 334–364). Hershey, PA: IGI Global. doi:10.4018/978-1-5225-0177-0.ch016

Wu, P., & Marek, M. (2016). Incorporating LINE Smartphone Affordances: Cross-Cultural Collaboration, Willingness to Communicate, and Language Learning. *International Journal of Computer-Assisted Language Learning and Teaching*, *6*(2), 56–73. doi:10.4018/IJCALLT.2016040104

Xiang, C. H., & Moon, D. (2018). "Current Affairs in Mandarin": A Student-Led TV Talk Show at LSE. In C. Xiang (Ed.), *Cases on Audio-Visual Media in Language Education* (pp. 118–154). Hershey, PA: IGI Global. doi:10.4018/978-1-5225-2724-4.ch006

Yang, P. (2017). English Journal Writing on Moodle for EFL College Learners: An Investigation of English Self-Efficacy and Writing Performance. In D. Tafazoli & M. Romero (Eds.), *Multiculturalism and Technology-Enhanced Language Learning* (pp. 13–32). Hershey, PA: IGI Global. doi:10.4018/978-1-5225-1882-2.ch002

Yeromin, M., & Charskykh, I. (2017). Use of the Codes of Audio-Visual Media: Strengthening the Sub-Lingual International Communication in Higher Education. In D. Tafazoli & M. Romero (Eds.), *Multiculturalism and Technology-Enhanced Language Learning* (pp. 116–128). Hershey, PA: IGI Global. doi:10.4018/978-1-5225-1882-2.ch007

Yeromin, M. B., & Charskykh, I. (2018). Universal and Specific Codes of Cultural Context in Audio-Visual Media: Collision and Mutual Enrichment in International Technology-Enhanced Language Learning. In D. Tafazoli, M. Gomez Parra, & C. Huertas-Abril (Eds.), *Cross-Cultural Perspectives on Technology-Enhanced Language Learning* (pp. 68–82). Hershey, PA: IGI Global. doi:10.4018/978-1-5225-5463-9.ch004

Zhang, S. (2016). A Blended Chinese-as-a-Foreign-Language Short Course: Design and Perceptions. *International Journal of Computer-Assisted Language Learning and Teaching*, *6*(2), 35–55. doi:10.4018/IJCALLT.2016040103

Zhao, K. (2016). Enhancing Metacognitive Language Learning Strategy Use and Business Language Proficiency in Technology-Enhanced Collaborative Learning Environment. *International Journal of Computer-Assisted Language Learning and Teaching*, *6*(3), 68–78. doi:10.4018/IJCALLT.2016070105

About the Contributors

Rashad Ahmed is a Visiting Assistant Professor at Miami University, Ohio. He completed his Ph.D. in Applied Linguistics at the University of Memphis, where he also earned a Graduate Certificate in TESL/ TEFL. He attained his master's degree in Applied Linguistics at Southern Illinois University Carbondale. He has been granted several awards, fellowships, and scholarships, which included his Fulbright scholarship 2013 to 2017, the Applied Linguistics Concentration Award and a Summer Doctoral Fellowship from the University of Memphis. His research interests involve: Computer Assisted Language Learning (CALL), Language Assessment, Digital Literacy, TESOL Methods, Sociolinguistics, and First-year Composition.

Abdu Al-kadi obtained his PhD (applied linguistics) from ISLT, Carthage University, Tunisia. He has contributed to undergraduate courses in English as a Foreign Language (EFL) and English for Specific Purposes (ESP) in Yemen. He has published on and continues to be agitated by issues within the realm of ELT, SLA, Netlinguistics, ESP, and CALL. He has authored, coauthored, and reviewed some articles published in peer-reviewed journals on these areas of research. He is also a member of the International Editorial Board of the Journal of Education and Science (EĞİTİM VE BİLİM) as well as Language Teaching and Educational Research (LATER) - both based in Turkey.

* * *

Mohialdeen Alotumi is Assistant Professor of TESOL and Applied Linguistics at the Department of English, Faculty of Languages, Sana'a University, Yemen. His research interests include CALL, EFL, ESL, MALL, TESOL, and Language Teacher Education.

Taha Hezam is an assistant professor of English at Taiz University for the last ten years. He is a well experienced educator.He is currently affiliated to Bisha University, KSA. He served as a lecturer, curriculum designer and administrator at the Faculty of Education, Taiz University. He worked as a lecturer since 1992. He was the head of the English Department for 5 years and he participated in many projects for developing the English Department's curriculum as well. He has an experience in teaching online courses through Blackboard platform that provides various technological facilities, such as virtual classrooms, discussion forum, wikies, multimedia recordings and online testing, etc.

Ferit Kılıçkaya is currently working as an associate professor at the department of English language teaching at Burdur Mehmet Akif Ersoy University, Turkey. His main area of interest includes computer-assisted language learning and testing, language teacher education, and language teaching methodology. He has published several book chapters, articles, and reviews in journals such as British Journal of Educational Technology, Computer Assisted Language Learning, Teaching English with Technology, and Educational Studies.

Saadet Korucu-Kis is an Assistant Professor in the English Language Teaching Department of Necmettin Erbakan University. She holds a BA, MA, and PhD in English Language Teaching. Her research interests relate to technology integration, teacher cognition, teacher technology education, and reflective practice.

Nor Rizan Maasum is an Associate Professor at the English Language Studies Program at the Faculty of Social Sciences and Humanities, Universiti Kebangsaan Malaysia. Her research interests include cooperative learning, second language reading, workplace and e-literacy.

Norizan Razak is a professor in digital literacy and e learning. She is currently the director of Tun Fatimah Hashim Women leadership center in Universiti Kebangsaan Malaysia. She has authored more than 100 journal articles and books related to computer assisted language learning. Currently completing research related to gender and workplace opportunities.

Emine Şendurur received her PhD in Computer Education and Instructional Technology from Middle East Technical University, Turkey, in 2012. In 2005, she was hired as a research assistant in the same department. After completing her PhD, she worked as an instructor in Computer Programming department. She recently works as an instructor in Computer Education and Instructional Technologies department at Ondokuz Mayıs University, Turkey. She teaches courses including instructional design, computer science, learning theories, Internet based programming and human-computer interaction. Her main research interests consist of user experience, informal learning, social networking sites, cognitive load theory, instructional message design, and eye-tracking methodology.

Amr Yassin is a lecturer at English Department, Center of Languages and Translation, Ibb University, Yemen. Currently, he is doing his PhD in English Language Studies at Universiti Kebangsaan Malaysia, Malaysia. He has published several papers related to English language teaching and learning. His areas of interest include ELT and E-learning.

Index

Purchase Print, E-Book, or Print + E-Book

IGI Global's reference books can now be purchased from three unique pricing formats: Print Only, E-Book Only, or Print + E-Book.
Shipping fees may apply.

www.igi-global.com

Recommended Reference Books

ISBN: 978-1-5225-9348-5
© 2019; 454 pp.
List Price: $255

ISBN: 978-1-5225-7763-8
© 2019; 253 pp.
List Price: $175

ISBN: 978-1-5225-7531-3
© 2019; 324 pp.
List Price: $185

ISBN: 978-1-5225-7802-4
© 2019; 423 pp.
List Price: $195

ISBN: 978-1-5225-6246-7
© 2019; 610 pp.
List Price: $275

ISBN: 978-1-5225-2998-9
© 2018; 352 pp.
List Price: $195

Looking for free content, product updates, news, and special offers?
Join IGI Global's mailing list today and start enjoying exclusive perks sent only to IGI Global members.
Add your name to the list at **www.igi-global.com/newsletters.**

Publisher of Peer-Reviewed, Timely, and Innovative Academic Research

IGI Global
DISSEMINATOR OF KNOWLEDGE

www.igi-global.com Sign up at www.igi-global.com/newsletters facebook.com/igiglobal twitter.com/igiglobal

Ensure Quality Research is Introduced to the Academic Community

Become an IGI Global Reviewer for Authored Book Projects

The overall success of an authored book project is dependent on quality and timely reviews.

In this competitive age of scholarly publishing, constructive and timely feedback significantly expedites the turnaround time of manuscripts from submission to acceptance, allowing the publication and discovery of forward-thinking research at a much more expeditious rate. Several IGI Global authored book projects are currently seeking highly-qualified experts in the field to fill vacancies on their respective editorial review boards:

Applications and Inquiries may be sent to:
development@igi-global.com

Applicants must have a doctorate (or an equivalent degree) as well as publishing and reviewing experience. Reviewers are asked to complete the open-ended evaluation questions with as much detail as possible in a timely, collegial, and constructive manner. All reviewers' tenures run for one-year terms on the editorial review boards and are expected to complete at least three reviews per term. Upon successful completion of this term, reviewers can be considered for an additional term.

If you have a colleague that may be interested in this opportunity,
we encourage you to share this information with them.

eContent Pro International
PARTNERING
IGI Global

IGI Global Proudly Partners With eContent Pro International

Receive a 25% Discount on all Editorial Services

Editorial Services

IGI Global expects all final manuscripts submitted for publication to be in their final form. This means they must be reviewed, revised, and professionally copy edited prior to their final submission. Not only does this support with accelerating the publication process, but it also ensures that the highest quality scholarly work can be disseminated.

English Language Copy Editing

Let eContent Pro International's expert copy editors perform edits on your manuscript to resolve spelling, punctuaion, grammar, syntax, flow, formatting issues and more.

Scientific and Scholarly Editing

Allow colleagues in your research area to examine the content of your manuscript and provide you with valuable feedback and suggestions before submission.

Figure, Table, Chart & Equation Conversions

Do you have poor quality figures? Do you need visual elements in your manuscript created or converted? A design expert can help!

Translation

Need your documjent translated into English? eContent Pro International's expert translators are fluent in English and more than 40 different languages.

Hear What Your Colleagues are Saying About Editorial Services Supported by IGI Global

"The service was very fast, very thorough, and very helpful in ensuring our chapter meets the criteria and requirements of the book's editors. I was quite impressed and happy with your service."

– Prof. Tom Brinthaupt,
Middle Tennessee State University, USA

"I found the work actually spectacular. The editing, formatting, and other checks were very thorough. The turnaround time was great as well. I will definitely use eContent Pro in the future."

– Nickanor Amwata, Lecturer,
University of Kurdistan Hawler, Iraq

"I was impressed that it was done timely, and wherever the content was not clear for the reader, the paper was improved with better readability for the audience."

– Prof. James Chilembwe,
Mzuzu University, Malawi

Email: customerservice@econtentpro.com

www.igi-global.com/editorial-service-partners

IGI Global
DISSEMINATOR OF KNOWLEDGE
www.igi-global.com

Celebrating Over 30 Years of Scholarly Knowledge Creation & Dissemination

InfoSci®-Books

A Database of Over 5,300+ Reference Books Containing Over 100,000+ Chapters Focusing on Emerging Research

InfoSci®-Books Database

The **InfoSci®-Books** database is a collection of over 5,300+ IGI Global single and multi-volume reference books, handbooks of research, and encyclopedias, encompassing groundbreaking research from prominent experts worldwide that span over 350+ topics in 11 core subject areas including business, computer science, education, science and engineering, social sciences and more.

Open Access Fee Waiver (Offset Model) Initiative

For any library that invests in IGI Global's InfoSci-Journals and/or InfoSci-Books databases, IGI Global will match the library's investment with a fund of equal value to go toward **subsidizing the OA article processing charges (APCs) for their students, faculty, and staff** at that institution when their work is submitted and accepted under OA into an IGI Global journal.*

INFOSCI® PLATFORM FEATURES

- No DRM
- No Set-Up or Maintenance Fees
- A Guarantee of No More Than a 5% Annual Increase
- Full-Text HTML and PDF Viewing Options
- Downloadable MARC Records
- Unlimited Simultaneous Access
- COUNTER 5 Compliant Reports
- Formatted Citations With Ability to Export to RefWorks and EasyBib
- No Embargo of Content (Research is Available Months in Advance of the Print Release)

**The fund will be offered on an annual basis and expire at the end of the subscription period. The fund would renew as the subscription is renewed for each year thereafter. The open access fees will be waived after the student, faculty, or staff's paper has been vetted and accepted into an IGI Global journal and the fund can only be used toward publishing OA in an IGI Global journal. Libraries in developing countries will have the match on their investment doubled.*

To Learn More or To Purchase This Database:
www.igi-global.com/infosci-books
eresources@igi-global.com • Toll Free: 1-866-342-6657 ext. 100 • Phone: 717-533-8845 x100

IGI Global
DISSEMINATOR OF KNOWLEDGE
www.igi-global.com

www.ingramcontent.com/pod-product-compliance
Lightning Source LLC
LaVergne TN
LVHW061534290926
849162LV00004B/791